Advances in
Computer Architecture

Advances in Computer Architecture

GLENFORD J. MYERS
IBM Systems Research Institute

A WILEY-INTERSCIENCE PUBLICATION

JOHN WILEY & SONS, New York • Chichester • Brisbane • Toronto

Copyright © 1978 by John Wiley & Sons, Inc.

All rights reserved. Published simultaneously in Canada.

Reproduction or translation of any part of this work beyond that permitted by Sections 107 or 108 of the 1976 United States Copyright Act without the permission of the copyright owner is unlawful. Requests for permission or further information should be addressed to the Permissions Department, John Wiley & Sons, Inc.

Library of Congress Cataloging in Publication Data:

Myers, Glenford J., 1946-
   Advances in computer architecture.

  "A Wiley-Interscience publication."
  Includes bibliographical references and index.
  1. Computer architecture. I. Title.
QA76.9.A73M93   621.3819'52   77-19001
ISBN 0-471-03475-4

Printed in the United States of America

10 9 8 7 6 5 4

TO MY PARENTS

Preface

Since the 1950s we have witnessed many advances in computing systems. The software field has advanced tremendously: for instance, we now have better tools, methodologies, and programming languages; software applications are more sophisticated; new algorithms have been invented; and the construction of such programs as operating systems and compilers is fairly well understood. The construction of physical computing devices has also advanced significantly: for example, circuit speeds and densities have increased by orders of magnitude, new storage technologies have been invented, better algorithms have been devised, and the microprogramming concept has been exploited. However, we have seen almost no advances at the hardware/software interface, the level of a system usually referred to as the *computer architecture*. To be fair, there have been some significant advances, but they have not received widespread attention and have not found their way into most conventional systems. For instance, if the instruction sets of most current large-scale systems, minicomputers, and microcomputers are examined, they will be found to be strikingly similar to those of machines designed in the 1950s.

The similarity of the architecture of today's systems to that of earlier systems can cause us to become complacent about the subject; we look around us and see tremendous software and hardware advances, but we also see that the architectures of current systems are virtually the same as those of earlier systems. Thus we might be inclined to assume that someone in the 1940s and 1950s invented all there is to be invented in the area of computer architecture. As a result, the architecture of future systems remains the same. This attitude motivated me to write this book: to destroy this complacency by showing that there are serious problems in current computer architectures, and to discuss advanced architectural concepts that will solve these problems.

The intent of this book can also be expressed by examining two possible alternative titles that were considered. One title was *Fifth-Generation Computer Architectures*. The term "fifth" was selected because of the feeling that the fourth generation is already on the drawing board and that these systems would undoubtedly retain the architecture of earlier systems. This title was discarded, however, because it looked too "flashy." Also, it would be misleading, because some of the concepts discussed in the book arose in certain second-generation systems. Another possible title was *Second-Era Computer Architectures*, but this title was discarded because of the feeling that prospective readers would confuse "era" with "generation" and form the impression that the book is a historical survey of the IBM 1401, Burroughs 200, and other "second-generation" (discrete transistor) machines.

The chapters in this book are organized into six parts. Part I defines computer architecture, takes a critical look at current architectures, and discusses a set of properties needed in future computer architectures. Parts II, III, IV, and V are case studies; they discuss four advanced architectures with many or all of the desirable properties discussed in Part I. Part VI discusses other aspects of computer architecture, such as input/output considerations and the optimization or "tuning" of an architecture.

The book is intended for two audiences: for use as a text in a "second course" on computer architecture (where the "first course" would presumably cover conventional architectural concepts), and to spread some of these ideas to computer professionals in general. The reader is expected to have a good grasp of computer system fundamentals. In particular, the reader should be knowledgeable of programming language concepts (e.g., the phrase "scope of names in a block-structured language" should be meaningful to the reader), have an understanding of the machine or assembly language of a conventional machine (e.g., S/370, PDP-10, CDC 6600), be familiar with operating system and compiler concepts (e.g., the term "reverse Polish notation" should be a familiar one), and have a grasp of the concept of microprogramming. A basic premise of this book is that this knowledge is prerequisite to the development of computer architectures.

I have found that the most effective way to understand an architecture is to do a mental compilation of a high-level-language program to the architecture; many of the examples in the book were developed

PREFACE

along this line. If the book is used as a text, the student should be assigned a number of small PL/I, COBOL, or FORTRAN programs to be mentally compiled to each architecture.

Finally, the opinions in the book are those of the author and do not necessarily represent the opinions, or future product directions, of the IBM Corporation.

GLENFORD J. MYERS

New York, New York
January 1978

Contents

PART I THE NEED FOR ARCHITECTURAL ADVANCES

1. A Definition of Computer Architecture 3

The Role of the Computer Architect, 6
References, 7
Exercises, 8

2. A Critique of the Conventional von Neumann Architecture 9

The Semantic Gap, 11
The von Neumann Architecture, 19
Other Undesirables, 21
References, 23
Exercises, 24

3. A Classification of Computer Architectures 25

Language-Directed Architectures, 26
Type-A High-Level-Language Architectures, 28
Type-B High-Level-Language Architectures, 29
Type-C High-Level-Language Architectures, 29
Application-Directed Architectures, 31
References, 33
Exercises, 36

4. Requisites for Improved Architectures   37

Self-Defining Data, 37
Self-Defining Data Objects, 43
Expression-Evaluation Stacks, 46
Subroutine Management, 49
Lexical-Level Addressing, 50
Capability-Based Addressing, 52
Variable-Size Storage Cells, 55
References, 55
Exercises, 56

PART II   A LANGUAGE-DIRECTED ARCHITECTURE

5. The Student-PL Machine   61

The Student-PL Language, 61
SPLM Storage Structure, 63
References, 67
Exercises, 68

6. Program Compilation and Execution on SPLM   69

Program Segments for IF Statements and DO Loops, 73
Subroutine-Call Example, 77
Significance of SPLM, 80
Exercises, 81

7. SPLM Instruction Set   83

Data-Access and Addressing Instructions, 84
Data-Operation Instructions, 86
Control Instructions, 88
Procedure Instructions, 91
Array-Storage Instructions, 93

PART III   A HIGH-LEVEL-LANGUAGE ARCHITECTURE

8. System Architecture of the SYMBOL System   97

System Configuration, 98
Job Flow Through the System, 101

CONTENTS

The SYMBOL Programming Language, 103
References, 107
Exercises, 108

9. Computer Architecture of the SYMBOL System — 110

Representation of Data, 110
The Name Table, 112
The Object-Code String, 117
Exercises, 124

10. SYMBOL Processor and Configuration Architecture — 125

The Main Bus, 125
Memory Management, 127
The System Supervisor, 136
The Central Processor, 141
The Translator, 144
The Remaining Processors, 145
Significance of the SYMBOL System, 145
Exercises, 146

PART IV A MULTIPLE-LANGUAGE-DIRECTED ARCHITECTURE

11. The Burroughs B1700 System — 151

B1700 System Architecture, 152
Implementation Considerations, 153
Storage and Performance, 154
References, 155

12. Burroughs B1700 COBOL/RPG Architecture — 156

Data Types, 157
Program Parameters, 157
Storage Structure, 158
Instruction Formats, 161
Machine Instructions, 163
Reference, 174
Exercises, 175

PART V A SOFTWARE-RELIABILITY-DIRECTED ARCHITECTURE

13. The SWARD Machine — 179

> Development of the Design Goals, 180
> Evaluation of Current Architectures, 183
> Development of the Architecture, 185
> References, 189

14. Program Compilation and Execution on SWARD — 190

> Data Types, 190
> The Module, 198
> Instruction Formats and Addressing, 202
> Fault Handling, 203
> Instruction Summary, 207
> A One-Module Example, 210
> A Two-Module Example, 216
> Significance of SWARD, 222
> Reference, 223
> Exercises, 223

15. SWARD Instruction Specifications — 225

> General Instructions, 226
> Arithmetic Instructions, 228
> Comparison Instructions, 230
> Boolean Instructions, 232
> String Instructions, 233
> Control Instructions, 234
> Addressing Instructions, 237
> Debugging Instructions, 240
> Calculation of the Address-Field Size, 242
> Internal Storage Objects, 244

PART VI RELATED TOPICS IN COMPUTER ARCHITECTURE

16. Input/Output Architecture — 249

> Front- and Back-End Processors, 250
> Associative-Storage Processors, 252

CONTENTS

The Relational Associative Processor, 255
The One-Level Store, 267
I/O in the SWARD Machine, 268
References, 271
Exercises, 272

17. Architecture Optimization and Tuning   273

Instruction-Set Optimizations, 274
Operation-Code Optimization, 279
Address Optimization, 284
References, 291
Exercises, 291

18. The Art of Computer Architecture   293

Conceptual Integrity, 293
Orthogonality, 295
Extensibility, 295
Implementation Freedom, 295
Technology Independence, 295
Formal Description, 296
Mental Compilation, 296
Language Validation, 296
References, 297

Answers to Exercises   299

Index   309

Advances in
Computer Architecture

I
The Need for Architectural Advances

1 | A Definition of Computer Architecture

Since the term "computer architecture" means different things to different people, the first order of business is defining precisely the meaning of the term as used in this book. Rather than beginning with the definition, however, it is worthwhile to begin by examining the overall design process of a computing system.

In general, we can define architecture as the distribution of function across a given level or boundary of a computing system and the precise definition of the boundary. That is, if we are establishing the architecture of a certain level of the system, the first step is determining which of the system's functions will be performed above the level and which will be performed beneath the level. Once this has been accomplished, the second step is developing a precise definition of the level's interfaces.

This implies that there are distinct types of architecture within a computing system. One way to define computer architecture is to distinguish it from other forms of architecture. Figure 1.1 is used to illustrate this distinction by viewing a system as a set of levels of abstraction. (One unfortunate problem associated with any conceptual picture such as Figure 1.1 is that it may imply that certain decisions have already been made; no such implications should be drawn.)

The first type of architecture, called *system architecture* and represented by the levels labeled 1, is the determination of which data-processing functions are to be provided by the system and which ones are the responsibility of the outside world (e.g., end user, end-use ap-

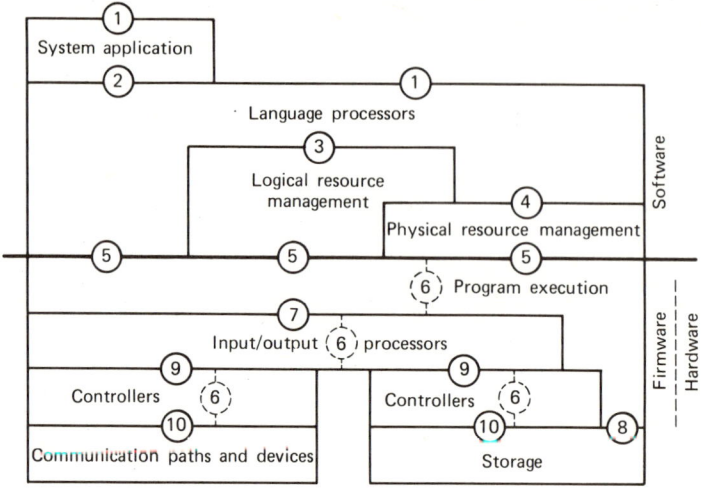

Figure 1.1 Levels of architecture within a computing system.

plications, system operator, data base administrator). The system probably presents itself to the outside world through two sets of interfaces: languages (e.g., terminal command language, programming languages, data base description and manipulation languages, operator language, job-control language) and system applications (application programs that are provided by the system developer, such as sorts, utilities, and information-retrieval programs). The system architecture process encompasses the definition of both sets of interfaces.

Interfaces 2, 3, and 4 represent distinct types of architectural processes within the system's software, although they are not generally recognized by any widely used names. If the system applications are written in languages not provided to the end user, another type of architecture is concerned with the definition of these languages. The language processors in turn see lower software levels of abstraction represented by architectural levels 3 and 4. The logical-resource management level might represent such functions as data base management, file management, virtual-storage management, and teleprocessing-network management. The physical-resource management level might represent such functions as the management of secondary and main storage space, the management of the underlying processors (i.e., process scheduling and synchronization), and the management of other physical devices. For lack of a better term, we can refer collectively to the determination of levels 2, 3, and 4 as *software architecture*.

A DEFINITION OF COMPUTER ARCHITECTURE

Level 5 represents a major dividing point; it is the boundary between the system's software and its firmware (microcode) and hardware. (The terms firmware/microcode/microprogram have no universally agreed on definition and are often misused. Their meaning here is the traditional one: a microprogram is a stored program that explicitly controls the data flow through the physical components, such as busses, registers, and adders, of a processor, and it is an alternative to performing data-flow control with a network of sequential logic circuits.) Hence level (interface) 5 represents the abstraction of the system's physical representation as seen by its software. Distributing function above and beneath this level and defining interface 5 is the process of *computer architecture*.

One can carry this idea further and talk about the distribution of function among parts of the physical system. For instance, interface 7 defines the amount of function to be performed by the program-execution processors (e.g., central processing units) versus input/output processors (e.g., channels). Another type of architecture is concerned with the distribution of function to the input/output processors versus the input/output device controllers. In turn, one must determine how function is distributed between the controllers and the physical input/output devices (e.g., terminals, modems, and disk and tape drives). These architectural levels (interfaces 7, 9, and 10) might be termed *physical input/output architecture*.

One remaining architectural level is number 8: the interface between the processor and main storage. The other remaining level is number 6. The function of each processor and controller might be distributed among microprograms, physical devices, and data paths. Hence the interfaces labeled 6 define the microprogram interface (e.g., the physical data flow and the microinstruction format) within each processor. Level 6 might be labeled *microprogram architecture*. Levels 6 and 8 are also commonly called *processor architecture* or *processor organization*.

One last form of architecture not explicitly shown in Figure 1.1 might be termed *configuration architecture*. (It is not shown because it involves vertical, rather than horizontal, cuts through the system.) Configuration architecture is concerned with the decision to distribute system function across multiple processors (e.g., deciding that the system will contain a "back-end" data base processor or that the system will be distributed among a network of minicomputers, and then distributing function across these processors and defining their interfaces).

Given this overall view of the types of architecture within a computing system, we can now define the term computer architecture. Computer architecture is a definition of a physical system (microcode and hardware) as seen by a machine-language programmer or a compiler writer. It is the definition of the conceptual structure and functional behavior of a processor as opposed to such factors as the processor's underlying data flow and controls, logic design, and circuit technology. Occasionally (e.g., [1]) some or all of these latter aspects are included in definitions of computer architecture, but they are excluded from the definition used in this book because they represent separate and distinct sets of decisions.

The computer architect, then, makes three broad classes of decisions: the form in which programs are presented to the underlying machine, the methods with which these programs name or address their data, and the representation of data. Within these classes the architect is faced with such decisions as determining the smallest addressable unit of storage, defining the types and formats of data, defining the machine instruction operations and formats, defining methods of storage addressing and protection, determining the sequencing mechanisms among instructions, and defining the view of input/output devices that is presented in the machine interface.

As an example of the distinction between computer architecture and the underlying levels of architecture, the intent of IBM's S/360 and S/370 family of computers was to provide a set of processors with a common computer architecture, but allowing each processor to have a different internal structure to meet certain price/performance objectives. With two exceptions, there is a single "principles of operation" manual [2] defining the computer architecture of all the S/370 processors. One exception is the few, but significant, architectural differences among the processors (e.g., the manner in which the processor informs the software of a machine error); these differences are defined in separate manuals for each processor. The second exception is that the computer architecture view of the I/O devices (in S/370 terminology, the definition of the channel command words) is not defined in this manual; it is defined in a separate manual associated with each device.

THE ROLE OF THE COMPUTER ARCHITECT

This definition of computer architecture gives us a clue to the computer architect's job. The major task of a computer architect is determining which system functions are best performed beneath level 5 of Figure 1.1 (i.e., by microcode and hardware) and which system functions are best performed above level 5 (i.e., by software), and defining the interface between them.

The major premise of this book is that most computer architects have not viewed their roles this way. One piece of evidence supporting this idea is our preoccupation with the term "MIPS" (millions of instructions processed per second) and the feeling that this is a reasonable measure of the speed of a computing system. We tend to take it for granted that an 8-MIPS machine is considerably faster than a 2-MIPS machine, but as we see in later chapters, quite the opposite might be true. For instance, if the computer must generate 21 instructions for the PL/I statement

(SUBSCRIPTRANGE):   A(I) = B(J);

on the eight-MIP architecture, but only one instruction on the two-MIP machine, it is no longer obvious that the eight-MIP machine is faster; in fact, from a system efficiency point of view, we may find it several times slower than the "slower" machine.

If the architect views the job of architecture properly, he or she should be concerned with the efficiency of problem solution, rather than the average raw speed of the machine instructions; that is, the architect must consider efficiency from the point of view of the programming-language/compiler/machine triplet. It is not the raw instruction speed that makes for an efficient system; rather, the "power" of the instruction set (the amount of function performed by the machine) and the number of bits that must be transmitted between the storage media and the processor to execute a given program have a more significant effect on performance. This implies that the computer architect must be driven primarily from the forces from above. The suggestion [3] that an architect for a new machine should have personally designed, coded, and debugged a compiler and operating system for a previous machine may be an unreachable ideal, but the thought is in the appropriate direction.

REFERENCES

1. S. S. Reddi and E. A. Feustel, "A Conceptual Framework for Computer Architecture," Computing Surveys, 8 (2), 277–300 (1976).
2. IBM System/370 Principles of Operation. GA22–7000, Poughkeepsie, N.Y.: IBM Corp., 1974.
3. W. M. McKeeman, "Language Directed Computer Design," Proceedings of the 1967 Fall Joint Computer Conference. Washington, D.C.: Thompson, 1967, pp. 413–417.

EXERCISES

1.1 Can a system's computer architecture be defined before its system architecture?

1.2 Given the following list of candidates—an application programmer, the compiler designer, the operating system designer, or the engineer designing the processor organization—which one is likely to perform best in specifying the computer architecture of a new system?

1.3 Examine the computer architecture of a system with which you are familiar. Do you see any evidence that the architecture was designed by careful consideration of hard/software tradeoffs, or does the system have the appearance of a "bottom-up" design?

2 | A Critique of the Conventional von Neumann Architecture

The main premise of this book is that the architectures of most current computing systems have not been designed according to the definition of computer architecture in Chapter 1. Rather than taking a global look at system function and its hardware/software tradeoffs, most computer architects have based their designs on tradition and the bottom-up view of "minimize the cost of hardware and let the programmers solve all the difficult problems."

One way of substantiating this statement is to show that, except for a few machines (e.g., some made by Burroughs Corporation), there have been no advances in the computer architectures of current systems since the 1950s. Some so-called advances that might come to mind (e.g., cache memories, instruction pipelining, the microprogramming concept) are not computer-architecture advances; they are advances in the implementation of particular architectures, that is, they are processor architecture or organization advances. In fact, some of these implementation advances can be viewed as *steps backward* in terms of computer architecture. For instance, the concept of pipelining introduced the architectural nuisance known as the imprecise interrupt [1]. Because a pipelining machine operates on multiple sequential instructions in parallel and may even execute independent instructions out of sequence, the machine cannot easily designate the precise instruction causing an error and cannot guarantee that instructions following the failing instruction have not been executed. Hence the programmers are given inaccurate information about their errors and are left to their own devices to figure out what went wrong.

If one compares the architectures of most current widely used machines (e.g., IBM System/370 and System/32, DEC PDP-10 and PDP-11, CDC 6600, Univac 1108, Intel 8080) to the EDVAC and EDSAC, the first electronic stored-program computers (built in the 1940s), all the significant differences will be found to have originated in the 1950s. These post-EDVAC differences are

1. *Index registers*—storage addresses are formed by adding the value of a designated register to a field in the instruction. This concept originated in 1949 in a University of Manchester computer. The Datatron computer (ElectroData Corp., 1953) was the first product to employ the concept.
2. *General-purpose registers*—removal of the distinction between index registers and accumulators and providing more than one accumulator register. It appears that this concept originated in the Pegasus computer (Ferranti, Ltd., 1956).
3. *Floating-point data representation*—representation of, and operations on, numbers expressed in exponent/mantissa form. This concept existed in 1954 in the IBM NORC and 704.
4. *Indirect addressing*—instructions may point to addresses which in turn point to the instruction's operand. This concept existed in 1958 in the IBM 709.
5. *Program interrupts*—instruction execution is diverted to a predefined location when an external event occurs. This concept originated in the Univac 1103 in 1954.
6. *Asynchronous input/output*—input/output operations are controlled by independent processors in parallel with normal instruction execution. The LARC computer (Remington Rand) contained a stored-program input/output channel (1956); it also contained two central processing units (multiprocessing).
7. *Virtual storage*—giving a system the appearance of having more main storage than it physically contains. The Atlas system (University of Manchester, 1959) introduced the concepts of paging and hardware dynamic address translation.

Although current systems differ significantly from their predecessors in terms of cost, speed, reliability, internal organization, and circuit technology, the computer architecture of most current systems has not advanced beyond the concepts of the 1950s. Current architectures have been driven largely by tradition and by underlying physical considerations. However, the architectural motivations of the 1940s and 1950s—trying to minimize hardware costs at the expense of software—make little sense in today's world of radically different computing-system economics.

THE SEMANTIC GAP

Since the second major premise of this book is that current computer architectures have serious shortcomings, it is worthwhile to examine these shortcomings before immediately jumping to conclusions about improvements. Most of these shortcomings are attributable to a phenomenon known as the *semantic gap* [2]. The semantic gap is a measure of the difference between the concepts in high-level languages and the concepts in the computer architecture. Most current systems have an undesirably large semantic gap in that the objects and operations reflected in their architectures are rarely closely related to the objects and operations provided in the programming languages. As we see later, this large semantic gap contributes to software unreliability, performance problems, excessive program size, compiler complexity, and distortions of the programming languages, all of which contribute negatively to the economics of data processing.

To understand the presence of the semantic gap, one can pick a programming language and a computer architecture and study the relationships between the two. As an example, we analyze PL/I and the IBM S/370. The example is not PL/I oriented, however, since most or all the PL/I concepts discussed also exist in such languages as COBOL, FORTRAN, and ALGOL. Neither is the example S/370 oriented; the S/370 was selected because it is representative of most conventional architectures and because there is a good chance that most readers are familiar with its architecture.

The following is a list of a few major and heavily used concepts in PL/I (or any other language for that matter). The question for each is determining to what S/370 (or most other architectures for that matter) concepts it is related.

1. *Arrays.* The array is the most frequently used language data structure. PL/I provides such concepts as multidimensional arrays, performing operations on entire arrays, referencing cross-sections (subarrays within arrays), and, hopefully, ensuring that array subscripts do not fall beyond the bounds of the corresponding array dimen-

sions. The question is, what S/370 concepts directly relate to these concepts? The answer is, absolutely nothing. The only architectural concept that seems indirectly related in a primitive way is the concept of index registers. This means that it is left to the compiler to create the widely used concept of an array out of the rather distant S/370 instruction set.

2. *Structures.* A second frequently used data concept is the structure, a collection of heterogeneous data elements (also known as a *record* in some programming languages). One finds absolutely nothing in the S/370 that is related to structures and operations performed on structures.

3. *String Processing.* The PL/I language contains the concepts of fixed and varying-size strings and string-processing operations such as concatenation, extracting a specified substring from a string, searching a string for a specified substring, computing the current length of a string, and checking whether the elements of one string occur at least once in another string. One finds nothing in the S/370 architecture that corresponds to these concepts. Of course, since PL/I compilers exist for the S/370, the concepts were implemented, but they are constructed by the compilers using rather primitive S/370 instructions. (In the earlier S/360 architecture, the compilers' task was even more difficult, since the S/360 instructions that were used could only operate on 256 bytes of storage, but the PL/I strings can contain up to 65534 bytes.) Handling of PL/I bit strings is further complicated by the fact that the S/370 can only address storage in 8-bit quantities (bytes).

4. *Procedures.* The basic program structure in PL/I is the procedure (subroutine). A procedure call entails saving the state of the calling procedure, dynamically allocating and initializing local storage for the called procedure, transmitting arguments, and beginning execution of the called procedure. One finds next to nothing in the S/370 that corresponds to these concepts. One tiny exception is the branch-and-link instruction, but this contributes so little to the procedure-call operation (one of many instructions that must be executed) that its absence would never be missed (the compiler could just as easily generate two instructions, load-address and branch-register, in its place).

5. *Block Structure.* PL/I is a block-structured language, implying, for one thing, the existence of scope-of-name rules that define the addressability of references in inner blocks to undeclared variables.

THE SEMANTIC GAP

There is nothing in the S/370 that corresponds to this addressing concept, again meaning that the concept must be created by compiler-generated code.

6. *ON-units.* PL/I contains a concept of program-oriented interrupts when exceptional or predefined conditions occur. The concept includes such ideas as defining which on-units will handle which conditions, dynamically enabling and disabling on-units, and determining the scope of on-units across block and procedure invocations. The only S/370 concept that is related to this concept is its interrupt mechanism. However, S/370 interrupts are systemwide (not local to a particular program), meaning that the operating system must get into the act. Also, most of the PL/I conditions (e.g., conversion, subscript-range, area, error, name, check) have no related S/370 interrupts, implying that the compiler must generate code to detect and manage these conditions.

7. *Data Representations.* PL/I has decimal and binary fixed-point data representations (integer.fraction). The S/370 has none, but it does have decimal and binary integer representations out of which the compiler must create the fixed-point concept. PL/I decimal numbers can contain anywhere from 1 to 15 digits, but the S/370 can only represent decimal numbers with an odd number of digits. PL/I binary numbers can contain anywhere from 1 to 31 binary digits, but the S/370 provides for only binary numbers of 15 or 31 digits. PL/I floating-point numbers can be declared as having 1 to 33 digits of significance, but these must be mapped into one of three fixed-size S/370 representations.

This discussion could be carried on indefinitely by looking at other PL/I concepts such as controlled storage (a push-down stack concept), generic procedure calls, program-tracing functions, and automatic data conversion, but by now the reader should have an understanding of the semantic gap between high-level-language concepts and current computer architectures. The cause of large semantic gaps is more difficult to discover, but the usual causes are bottom-up system design and the computer architect's lack of knowledge and appreciation of programming languages, what programs do, what programmers do, the difficulty of program debugging, and the causes and consequences of software errors.

Given the existence of this large semantic gap, the next step is to discuss some of its consequences.

Software Unreliability

The semantic gap is a significant contributor to software unreliability in the sense that a large set of programming errors that could theoretically be prevented or detected by the computing system are not prevented or detected in current systems. A few examples suffice here, since this matter is discussed in more detail in later chapters, particularly in Part V.

One common programming error that arises under a large variety of circumstances is a reference to a variable that has an undefined or unset value. This error is not detected by most current systems; since execution proceeds using some unpredictable value, the error is difficult to debug. The reason that the error is not detected is that, although some instances of it could be detected at compilation time by doing a flow analysis of the program, in general it cannot be detected until execution time. Since conventional machines have no way of distinguishing a defined variable from an undefined one, the architects have, in effect, thrown the problem over the wall to the compiler writer. The compiler writer finds no easy and efficient solution to the problem; thus he or she throws the problem over the wall to the application programmer.

Some compilers have attempted to solve the problem, but the solutions have turned out to be complicated, inefficient, and not foolproof. For instance, IBM's PL/I Checkout Compiler initializes all character strings with hexadecimal FE characters and all fixed-point binary numbers with the smallest negative number and then checks for these values whenever these variables are referenced. However, not only does this add overhead (execution time and storage), but it can cause "errors" to be detected in correct programs and does not cover all data types. It is shown later that this error could be detected at virtually no cost by the appropriate computer architecture.

A second common error is referencing an array element where one of the subscripts falls beyond the bounds of the corresponding dimension. Again, since the conventional machine knows nothing of arrays, the problem is presented to the compiler writer. The compiler writers see no easy solution, thus they either ignore the problem or leave the decision to the application programmer by making the check optional. However, since the optional check is likely to add overhead, the compiler writer also adds a warning in the language manual such as, "Since this checking involves a substantial overhead in both storage space and execution time, it usually is used only in program testing—it is removed for production programs" [3]. Such a strategy is unwise because it "is like a sailor who wears his life preserver while training

THE SEMANTIC GAP

on land but leaves it behind when he sails" [4]. In other words, it seems peculiar to perform error checks while testing (when an erroneous result causes no harm) and then remove them in production runs (when an erroneous result could be disastrous).

As an example of the overhead of this software check, IBM's PL/I Optimizing Compiler normally generates 17 machine instructions (occupying 62 bytes of storage) for the statement

C(I,J) = A(I,J) + B(J,I);

when A, B, and C are arrays of fixed-binary elements of identical size. If the optional SUBSCRIPTRANGE check is enabled, the compiler generates 75 machine instructions (274 bytes), and 57 of these instructions would be executed if the subscripts were within the array bounds. Hence the optional check increases the object-code storage space for this statement by 340% and its execution time by a factor of about 3. It is shown later that this check could be performed by the machine with absolutely no overhead. ("No overhead" actually means "negative overhead"; a machine that performs this check should actually run faster than a conventional machine with software that does not perform the check!)

In his book [5] on the ethics of computing, even Weizenbaum, in describing in laymen's terms how computers work, seems to allude to the existence of the semantic gap and its relationship to reliability (the material in brackets is mine):

Computers are maddeningly efficient at stumbling over purely technical, i.e., linguistic, programming errors, but stumbling in a way that disguises the real locus of the trouble.... A real reason that programming is very hard is that, in most instances, the computer knows nothing of those aspects of the real world that its program is intended to deal with ... [i.e., the fact that "2.999 ... people" cannot exist]. It is in fact very hard to explain anything [i.e., language concepts] in terms of a primitive vocabulary [i.e., the machine instructions] that has nothing whatever to do with that which has to be explained. ... To write a good sonnet or a good program, one must know what one wants to say. And it helps enormously if one's critic [the machine] shares one's relevant knowledge base [the language concepts].

Performance Problems

The large semantic gap also leads to significant performance problems because of the large number of instructions that must be generated by the compiler to implement the language concepts out of the rather primitive machine-instruction repertoire. This has a negative effect on performance because it increases the amount of information that must be transmitted between storage and the processor, and this has been found to be a good first-order measure in comparing the performance of different machines.

Since this effect is not widely understood, it is worthwhile to look at a simple example. Assume that we wish to add two 100 by 100 element fixed-binary PL/I arrays together. Hopefully we would write this as A=A+B; (writing nested DO loops to accomplish this is much more inefficient). IBM's S/370 PL/I Optimizing Compiler generates efficient object code for this statement: six instructions followed by a loop of four instructions executed 10,000 times. The number of 32-bit words that must move between memory and the processor is 40,004 (the instructions; the first six instructions fit into four words, and the loop body occupies four words) plus 30,003 (two data fetches and one store for the elements plus a few additional fetches), for a total of 70,007. If we were compiling this to a machine that recognizes arrays (e.g., the machine in Part V), only one machine instruction might be needed (add B to A), and the memory/processor word transfer would be in the neighborhood of 30,000 (30,000 plus one for the instruction plus perhaps a few more for descriptive information about the arrays). Also, where the first machine had to decode and interpret 40,006 instructions, the second machine only decodes one instruction. Certainly there are execution factors to consider other than memory/processor transfer and instruction decoding (i.e., address calculation and the addition process), but these factors should be approximately the same in both machines.

Although this example applies only to array operations, one can find analogous examples in the excessive number of instructions generated to implement almost every programming-language concept on a conventional architecture.

Excessive Program Size

The large semantic gap affects program size in the same way. For instance, we saw earlier that it takes 62 bytes of storage to represent the statement

C(I,J) = A (I,J) + B(J,I);

if no subscript checking is done and 274 bytes if subscript checking is desired. In contrast, on the machine described in Part V, this statement can be expressed in two machine instructions with a total size of 13 bytes (and subscript checking is automatically performed).

In addition to being a problem itself, excessive program size is another contributing factor to system performance problems (e.g., in a virtual-storage system, by increasing the programs' working-set sizes and thus increasing the number of page faults incurred).

Compiler Complexity

From the previous two points, the effect of the large semantic gap on compilers should be obvious; the code-generation portion of compilers must be extremely complex to generate code that bridges the semantic gap as efficiently as possible.

One common reaction to this, particularly from the engineer designing the processor, is, "So what? The complexity must exist somewhere, so you are simply shifting it from one place (the compiler) to another (the machine, probably its microprogram)." From a processor-cost point of view, this comment may have merit, but it is not a valid argument in most computing systems. If we assume that the system has more than one compiler (hundreds have been written for the S/370), the semantic gap must either be bridged in every compiler or just once in the machine's implementation. For instance, if the programming languages provided have concepts in common such as array subscripting and procedure calls, it makes more sense (even disregarding the other advantages of doing so) to bridge these gaps once in the machine's architecture rather than in each compiler.

Programming Language Distortions and Misuse

If the semantic gap is so large that the compiler cannot efficiently bridge all of it, the consequences are that the definition of the language is distorted, the underlying machine shows through the language, and the language is misused.

As an illustration, assume that a S/370 PL/I programmer declares a variable A as a fixed-point decimal number of two digits. One would expect the assignment statement A=100 to fail, but it does not, and if variable A was printed, it would have the value 100 because the S/370 can only represent decimal numbers with an odd number of digits. Seeing that a complete bridge of this semantic gap would lead to inefficient object code, the compiler designers chose to map two-digit PL/I decimal variables into three-digit S/370 operands. The other possible alternative, changing the PL/I language definition, would be equally confusing to programmers and would make the language excessively S/370-architecture oriented.

As another illustration, PL/I contains both decimal and binary number representations. It is difficult to see why any PL/I programmer would use binary representations, for if one has a fixed-binary variable M and wants to assign it the value 25, the proper assignment statement is

M = 1101B;

(Most PL/I programmers get sloppy and write M=25, but "25" is a decimal number, causing PL/I to invoke its automatic data conversion rules, and these rules can lead the uninitiated into many strange errors [6].) However, many S/370 PL/I programmers use binary variables because they know that the S/370 is primarily a binary machine and that decimal numbers take more storage and decimal operations have a considerably longer execution time. Therefore the underlying computer architecture causes the programming language to be misused.

As a final PL/I example, the language contains the concept of bit-string variables. However, the unsuspecting programmer who attempts to pass a bit-string argument to another procedure will often encounter an unexpected result. Since the S/370 can only address storage to byte boundaries and since bit strings do not necessarily start on byte boundaries, there may be storage alignment problems between the two procedures. (The reader experienced in PL/I may recognize that the ALIGNED attribute was added to PL/I to correct this type of problem, but this is just another sign of the underlying machine influencing the language.)

One can find distortions in other languages. For instance, the FORTRAN arithmetic IF statement branches to one of three other statements depending on whether the predicate expression is negative, zero, or greater than zero. This came about not from language-design aesthetics, but because the IBM 704 computer had a compare instruction that branched to one of three succeeding instructions. Another peculiarity in FORTRAN is that DO loops always execute at least one iteration. Again, this decision came from the original underlying machine: the use of the TIX instruction on the 704.

In summary, the large semantic gap in current systems has many significant shortcomings, and these in turn have a negative effect on the economics of the development and use of computing systems. Although most of the illustrations were taken from PL/I and the S/370, they apply to most programming languages on most conventional computer architectures.

THE VON NEUMANN ARCHITECTURE

The basic reason for the existence of the large semantic gap in current systems is that most architectures are simply modifications of the von Neumann architecture model[7] derived in the 1940s. This is not to imply that the von Neumann architecture was not a stroke of genius when it was developed. However, the world has changed tremendously since the 1940s. The feasibility of even constructing electronic computers was still in doubt at that time, and hardware costs and reliability were of utmost concern; thus the motivation was to design as primitive a processor as possible. Also, factors that are taken for granted today, such as high-level programming languages and the sophistication and critical nature of most computing applications, were not even envisioned at that time.

It is common today to talk of a class of machines as von Neumann machines and to say that most current machines belong to this class. A von Neumann machine is said to have these properties:

1. A single sequential memory. A program and its data are stored in a single memory and the memory is referenced with sequential (0, 1, 2, ...) addresses.
2. A linear memory. The memory is one-dimensional, that is, it has the appearance of a vector of words.
3. No explicit distinctions between instructions and data. One can, for instance, treat an instruction as data (e.g., modify it), add an instruction to a data word, or branch to a data word and execute it as if its bits represent an instruction.
4. Meaning is not an inherent part of data. There is nothing, for instance, that explicitly distinguishes a set of bits representing a floating-point number from a set of bits representing a character string. Rather, the meaning of data is assigned by program logic. If a machine fetches a floating-point add instruction, it assumes that the operands represent floating-point numbers and performs a floating-point addition with the operands. Hence one can perform a floating-point addition on two operands that actually represent a character string and an address.

Although the von Neumann architecture was a reasonable design for the first stored-program computer, it is alien to the execution of programs written in high-level languages. In contrast to the four characteristics above:

1. Storage, as presented in a high-level language, consists of a set of discrete named variables. With the exclusion of certain questionable language constructs (e.g., the FORTRAN COMMON area), there is no concept of one variable being "next" to another variable. There is no reason to believe that the variables in one subroutine are located in the same storage device as the variables in another subroutine. In short, the concept of a single sequential storage bears little resemblence to the concept of storage in programming languages.
2. Programming languages deal with multidimensional, not just linear, data types (e.g., arrays, structures, and lists).
3. In programming languages there is a sharp distinction between data and instructions. In a high-level language, there are no concepts of executing data or referencing instructions as if they were data.
4. In a high-level language, meaning is an inherent part of data. One does not write a PL/I program as

    DECLARE A WORD;
    DECLARE B WORD;
    A = A "floating-point add with" B;

    Instead one writes

    DECLARE A DECIMAL FLOAT (6);
    DECLARE B DECIMAL FLOAT (6);
    A = A + B;

    That is, in high-level languages the meaning of the data is associated with the data itself, and the operators are generic (i.e., the meaning of + is determined by examining the attributes of its operands).

Thus we see that the attributes of a von Neumann architecture are not related, and are even contradictory, to the concepts in languages. A von Neumann machine is therefore a poor vehicle for the execution of high-level-language programs because

1. Excessive mapping is required in software (i.e., by the compiler in the form of compiler-generated code) to match the language concepts to the von Neumann view of storage. This has been referred to

as "absorbing the structure [of the data] into the logic of the program" [8]. This should be apparent to anyone who has examined the output of a compiler; the amount of code generated by the compiler to map the language concepts of storage and data to the underlying architecture usually greatly outweighs the amount of problem-solving code generated.
2. A von Neumann machine is excessively overgeneral (e.g., one can use a word that has no currently defined value, address anything in storage, add a character string to an instruction); since this generality is fortunately absent from programming languages, the compiler (and its generated code) is left with the task of removing the generality and ensuring that it does not interfere with the definition of the language.
3. Because the concept of storage in a von Neumann machine is rather primitive, the operations (instruction set) performed by the machine are constrained to be equally primitive.

Thus most of the architectural advances discussed in this book to alleviate the semantic-gap problem are associated with architectures that contain none of the four characteristics of the traditional von Neumann architecture.

OTHER UNDESIRABLES

Although the von Neumann model is the major cause of the large semantic gap, there are additional undersirable architectural properties of current systems that contribute to the gap.

Binary (Base Two) Arithmetic

In current machines, binary arithmetic is treated as almost sacred, but it almost goes without saying that humans find base-two arithmetic quite distasteful. Since proposals for decimal arithmetic often evoke emotional arguments, it is worth exploring the traditional arguments for and against decimal arithmetic.

Two arguments may be presented in favor of decimal arithmetic. First, since today's computing environment is highly input/output oriented and since few, if any, people would consider forcing human beings to communicate with computers in base-two terms, current systems waste an enormous amount of time performing conversions between decimal and binary representations. Second, the fact that a machine represents numbers in base-two form cannot be completely hidden from the human, since, for instance, most rational decimal fractions are represented as infinite-digit base-two fractions. This means that finite-length base-two numbers are often approximations of decimal numbers, a source of programming difficulty, programming errors, and confusing language definitions.

The traditional arguments against decimal arithmetic are that it is slower than binary arithmetic and that binary numbers can be stored more compactly than decimal numbers.

The two arguments against decimal arithmetic are subject to question. First, one must weigh the speed of the arithmetic algorithms against the overhead of converting decimal numbers to binary and back again. Second, decimal arithmetic circuits have been devised (e.g.,[9]) that are competitive with binary circuits in terms of speed and only slightly less competitive in terms of cost.

The second argument (space) has some merit, but it is not insurmountable. For instance, the machine described in Part V is a decimal machine, but, on the average, decimal numbers stored within it occupy fewer bits of storage than do binary numbers on conventional machines. Also, there are ways of encoding decimal numbers such that the differences in storage space are reduced (see the exercises at the end of this chapter).

Fixed-Size Storage Words

In an architecture with fixed-size storage words, deciding on the word size is probably the most difficult tradeoff facing the architect. If the word size is too small, the maximum value of the numbers that can be represented is too small, fractional (e.g., floating-point) numbers are excessively imprecise, and larger addresses are needed. On the other hand, larger words tend to waste storage, because studies of the distributions of data values in programs (e.g., [10]) indicate that values are not uniformly distributed; they are heavily skewed in favor of small values (e.g., the values zero and one are common, the values in the range 10–20 are more common than values in the range 59,470–59,480). Hence large words waste storage because their high-order bits or digits are likely to be zero.

The second problem with fixed-size words is that many languages (e.g., PL/I and COBOL) allow the programmer to declare the size of each variable, and the possible sizes usually vary over a large range (e.g., a PL/I decimal floating-point variable can be declared as having anywhere from 1 to 33 mantissa digits). If the compiler is able to accurately map this concept into a fixed-size-word machine, performance problems (excessive generated code) are a likely consequence. If the compiler designer decides that the concept of variable-size data cannot be efficiently and accurately mapped into fixed-size words, the underlying machine architecture shows through and distorts the language. Of course, one might argue that languages should not contain this concept, but the argument has little validity. The concept assists one in defining machine-independent languages, allowing programs to be transferred from one machine type to another. Also, if one has a PL/I variable called PERSONWEIGHT, it is desirable to declare it with the smallest possible meaningful size (i.e., FIXED DECIMAL(3)), allowing the compiler or the run-time environment to detect such errors as assigning a person a weight of 1732 pounds.

Registers

Another concept that is alien to the concepts in programming languages is the presence of program-addressable registers (e.g., the concept of general-purpose registers in the S/370). If the machine requires the use of registers for all arithmetic operations and if the number of registers is small (both are the case in most machines), the compiler is left with the task of generating code to manage the registers and optimize their use. This generated code is extraneous in that it contributes nothing toward the expression of the source program's logic. This is another example of the architect throwing problems over the wall by saying "Here are the registers, but you figure out how to manage and use them."

REFERENCES

1. D.W. Anderson, F.J. Sparacio, and R.M. Tomasulo, "The IBM System /360 Model 91: Machine Philosophy and Instruction Handling," *IBM Journal of Research and Development*, **11**(1), 8–24 (1967).
2. U. O. Gagliardi, "Report of Workshop 4—Software-Related Advances in Computer Hardware," *Proceedings of a Symposium on the High Cost of Software*. Menlo Park, Calif.: Stanford Research Institute, 1973, pp. 99–120.
3. *OS PL/I Checkout and Optimizing Compilers: Language Reference Manual.* SC33-0009. White Plains, N.Y.: IBM Corp., 1972
4. D. E. Knuth, "Structured Programming with GO TO Statements," *Computing Surveys*, **6**(4), 261–301 (1974).
5. J. Weizenbaum, *Computer Power and Human Reason: From Judgment to Calculation.* San Francisco: Freeman, 1976.

6. G. J. Myers, *Software Reliability: Principles and Practices.* New York: Wiley-Interscience, 1976.
7. A. W. Burks, H. H. Goldstine, and J. von Neumann, "Preliminary Discussion of the Logical Design of an Electronic Computing Instrument, Part I, Volume I," Princeton, N.J.: Institute for Advanced Study, 1946.
8. J. K. Iliffe, *Basic Machine Principles,* 2nd ed. London: Macdonald, 1972.
9. M. S. Schmookler and A. Weinberger, "High Speed Decimal Addition," *IEEE Transactions on Computers,* **C-20**(8), 862–866 (1971).
10. W. G. Alexander and D. B. Wortman,"Static and Dynamic Characteristics of XPL Programs," Computer, 8(11), 41–46 (1975).

EXERCISES

2.1. Although the message of this chapter is that the semantic gap must be reduced, there are practical limits on its reduction. What do you expect is the major problem encountered if the semantic gap is reduced too much?

2.2 Although architectural concepts to reduce the semantic gap have not yet been introduced, review the four characteristics of a von Neumann machine and list a related characteristic for each that might shrink the semantic gap.

2.3 Why is 1.0 rarely equal to 10.0 x 0.1 in most languages on current machines?

2.4 If decimal digits are encoded into four bits, how much less efficient is this than a base-two representation?

2.5 Can you find a more efficient encoding scheme for decimal numbers?

3 | A Classification of Computer Architectures

Before plunging into a detailed analysis of how the problems discussed in the previous chapter might be overcome, it is worthwhile to survey previous attempts to close the semantic gap and to place these attempts into various categories based on the strategy used. This chapter describes five basic categories of strategies to close the semantic gap and includes references of most of the significant previous efforts in each category. Since the semantic gap was discussed in the last chapter with respect to current high-level procedural programming languages, special-purpose architectures (e.g., machines devoted to array processing, signal processing, and Fast Fourier Transforms) are excluded; our attention is focused on "general-purpose" machines. (General purpose, as used here, implies a machine that is usable for a large variety of applications.) One should also note that most of the efforts mentioned are "paper" architectures: architectures that have not been implemented in the form of a working model.

Figure 3.1 can be used to introduce some of the basic approaches to computer architecture. The top path represents the traditional approach: an extensive compilation process translates a high-level-language program to a low-level machine-language program; the latter is then interpreted by the machine. The first category of semantic-gap-closing architectures is represented by the second path; the source program is compiled into a higher-level machine language which is in turn interpreted by the machine. Such architectures are known as *language-directed architectures*.

A CLASSIFICATION OF COMPUTER ARCHITECTURES

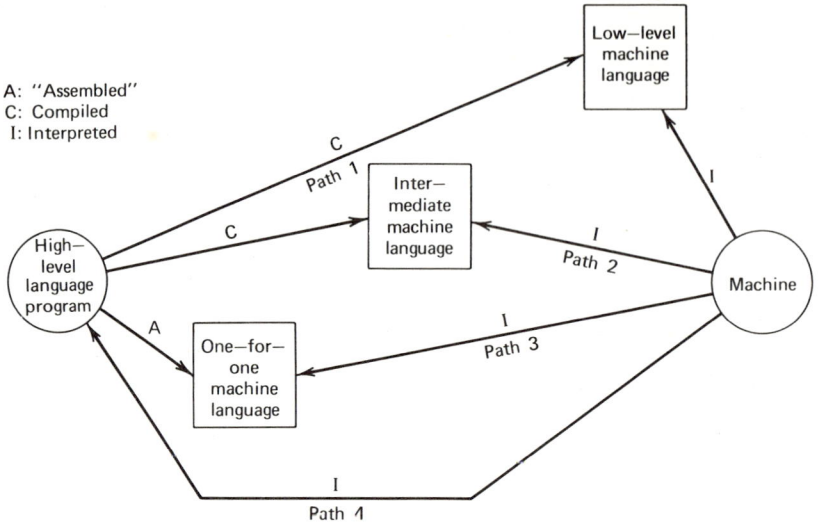

Figure 3.1 General language/machine relationships.

The last two paths represent three categories of architectures; machines in these categories are commonly called *high-level-language machines*. In path 3, the machine architecture is raised to such a level that the high-level language can be thought of as the assembly language; that is, there is a one-to-one correspondence between statement types and operators in the high-level language and instructions in the machine's instruction set. The source program is "assembled," rather than compiled, to the machine interface. This path represents two categories of architecture; the category is dependent on whether the assembly process is performed by software or by the machine. The other category is represented by path 4. In this case the high-level language is also the machine language in that the machine directly interprets high-level-language programs.

The fifth category (excluding the traditional approach) could be termed *application-directed architectures*. This category is not shown in Figure 3.1, but it represents general-purpose architectures that are oriented toward the efficient execution of a particular type pf program. The program usually selected is the system's operating system.

LANGUAGE-DIRECTED ARCHITECTURES

A language-directed architecture is one in which the operations and data structures implemented by the machine are more closely related to the operations and data structures in one or more high-level languages. However, as is the case with most terms in the computing field, this term has no precise or universally agreed on definition. For instance, if one added an instruction to the IBM S/370 that was semantically similar to the COBOL PERFORM-UNTIL statement, the S/370 would be slightly more "language directed," but not enough to fit within the spirit of a language-directed architecture. Rather, a language-directed architecture implies a machine that was designed with one or more high-level languages in mind, and this thinking permeates all architectural decisions.

The phrase "language-directed" appears to have been coined by McKeeman [1]; several others have discussed this type of architecture in a general way [2,3]. Architectures oriented toward ALGOL include the Burroughs B5500/6500/6700/7600 [4] and the KDF.9 [5]. There have also been numerous efforts to orient architectures toward the general properties of block-structured languages [2,6–9]. Another machine is oriented toward TPL, a block-structured language with APL-like operators [10].

The L-Machine is an architecture with a conventional storage architecture but with an instruction set oriented toward PL/I [11,12]. The architecture was implemented on a microprogrammed Microdata 1621. Performance measurements of the L-Machine show it to be superior to the execution of PL/I programs on a conventional architecture. Another PL/I-directed machine is that described by Sugimoto [13]. Also, the Student-PL Machine described in Part II of this book is oriented toward a subset dialect of PL/I [14].

A commercial machine directed toward COBOL is the Burroughs B3500, and there has also been at least one experimental attempt to direct an architecture toward COBOL [15]. APL has been a popular target for language-directed and high-level-language machines. The Raytheon AADC machine [16,17] is directed toward APL, as is at least one experimental architecture [18]. Other architectures have been oriented toward the BALM language [19], the MARY language [20], SNOBOL [21], and an unspecified algorithmic language [22].

Another approach in language-directed architectures is to orient the architecture toward the common semantics of a set of languages rather than toward one specific language. The Rice Research Computer [23], the BLM [24], and others [25] are examples of this approach. IBM Federal Systems Division's Intermediate Language Machine is oriented toward FORTRAN, APL, JOVIAL, and other languages [26,27]. The motivation behind the SWARD machine [28] (discussed in Part V) was not to produce a language-directed machine, but to produce a machine that significantly enhances the reliability of its programs, although the outcome was a machine that is directed toward PL/I, FORTRAN, COBOL, and other languages.

A machine that bridges the gap between these two approaches (specific versus general language directed) is the Burroughs B1700 machine discussed in Part IV. The machine's instruction set can be dynamically switched during operation (by invoking different microprograms), allowing the instruction set to be directed toward the language in which the program about to be dispatched was written.

TYPE-A HIGH-LEVEL-LANGUAGE ARCHITECTURES

The next architectural approach is represented by the third path in Figure 3.1. The architecture is directed toward the language to such an extent that the high-level language can be thought of as the assembly language (symbolic machine language) of the system. That is, there is a one-to-one correspondence between the statement types and operators in the high-level language and the machine operations. The program that translates source programs into machine-language programs (which we normally refer to as a compiler) is more accurately termed an assembler. The assembler strips comments and blanks from the program, translates operators, delimiters, and keywords into shorter machine-recognizable codes, converts names to addresses, and possibly rearranges expressions into reverse Polish notation, but it performs no traditional compilation functions (i.e., the code-generation process is trivial).

Although this approach is considered the second non-von Neumann architecture category, there is no concise distinction between this approach and the language-directed approach. Rather, the distinction is one of degree; one could consider the type-A high-level-language architecture as a highly language-directed architecture.

Examples of architectures in this category include a real-time system for the FLUID language [29], an aerospace system for the Space Programming Language (a derivative of JOVIAL) [30], a FORTRAN machine [31], an ALGOL-W machine [32], a system for the ISPL language [33], and a APL machine [34]. In addition, an IBM S/360 Model 25 was remicroprogrammed to generate an APL machine [35], and the APL Assist feature on the IBM S/370 Models 135, 138, 145, and 148 provides an additional microprogram to allow these machines to interpret "assembled" APL programs [36]. Also, the internal organization of the Burroughs Interpreter processor [37] was designed to facilitate the development of type-A high-level-language architectures.

TYPE-B HIGH-LEVEL-LANGUAGE ARCHITECTURES

This architectural category is also represented by path three in Figure 3.1 and is almost identical to the type-A architecture. The only difference is that the assembly process is performed by software in a type-A machine, but it is performed by the machine in the type-B approach. That is, the machine assembles the source program and then interprets the assembled program.

Examples of type-B machines include APL machines [38–40], a FORTRAN machine [41], a EULER machine [42], a syntax-driven machine [43], an ALGOL machine [44], and the SYMBOL system discussed in Part 3 of this book [45].

The type-B machine has the same semantic gap as a type-A machine, but the disadvantages of a type-B machine seem to outweigh its advantages. Its only advantage over a type-A machine is that the assembly process should be faster because it is implemented as a microprogram or in hardware; however, since source programs are assembled rather than compiled in both approaches, this difference should prove to be insignificant in most environments. Its disadvantages are higher cost and lower extensibility, because it is generally recognized that it is more expensive to implement a given function in microcode or hardware than in software and that functions implemented in microcode or hardware are more costly to change. Architects often use the following three criteria in determining whether a function should be implemented in the machine rather than in software: (1) the function should be small, (2) the function should be unlikely to change, and (3) system performance would suffer from a slower software implementation of the function. It is not clear that the assembler function meets *any* of these criteria.

TYPE-C HIGH-LEVEL-LANGUAGE ARCHITECTURES

This extreme approach to computer architecture is represented by the bottom path in Figure 3.1. In such a machine, the machine language is identical to the high-level language; that is, the machine directly interprets high-level-language programs.

Examples of such architectures include an APL machine [46], ALGOL 60 machines [47,48], an ADAM-language machine [49], an IPL-language machine [50], a machine for a string and list-processing language [51], and a machine whose structure can be tailored to any language [52]. Readers familiar with the IBM 5100 portable computer might consider it to be in this category, but it is not. When the 5100 is interpreting APL programs, its processor has an architecture that is a subset of the S/370 architecture. The processor interprets a software APL interpreter, which in turn interprets the APL program. The same holds when the 5100 is interpreting BASIC programs, but in this case the processor is loaded with an instruction set similar to the IBM System/3 and interprets a software BASIC interpreter.

Although the type-C architecture reduces the semantic gap to zero, its disadvantages seem so overwhelming that such architectures will probably never rise beyond the "curiosity" level. First, it is not clear whether the art of computer architecture has been performed at all in this approach. Computer architecture is a tradeoff process, determining which functions are best implemented by software or in the underlying machine; however, in this approach, everything has been relegated to the machine.

Type-C architectures must also be less efficient than the previous approaches, because every time the machine executes a statement, it must perform a lexical analysis of the statement, parse the statement, convert symbols to addresses, and so on. This must be done each time a statement is executed (e.g., if a particular IF statement is executed 1000 times, it is parsed 1000 times), but in the other approaches, this would only be done once per statement by the compiler or assembler.

For the same reasons as in the type-B approach, this architecture is more costly and less extensible than the other categories. In addition to the lower development cost of implementing a function in software rather than in microcode or hardware, a distinct advantage of software is that it has a negligible manufacturing cost. A microprogram also has a negligible manufacturing cost, but since microprograms are normally stored in a higher-speed and more costly memory (a control storage), a system with an excessive amount of microcode has a higher manufacturing cost than a system in which some of this function is implemented in software. Needless to say, implementing a function in hardware (i.e., as a sequential-logic network) results in higher development and manufacturing costs.

These factors are illustrated in Figure 3.2, which represents an educated guess of the expected system cost/performance ratio of the computer architecture approaches. The cost/performance ratio represents the system as a whole; that is, it is meant to include the performance and development, manufacturing, and servicing costs of the system's software and the underlying machine. The type-C architectural approach might be expected to have a *higher* cost/performance ratio than even the conventional von Neumann architecture.

APPLICATION-DIRECTED ARCHITECTURES

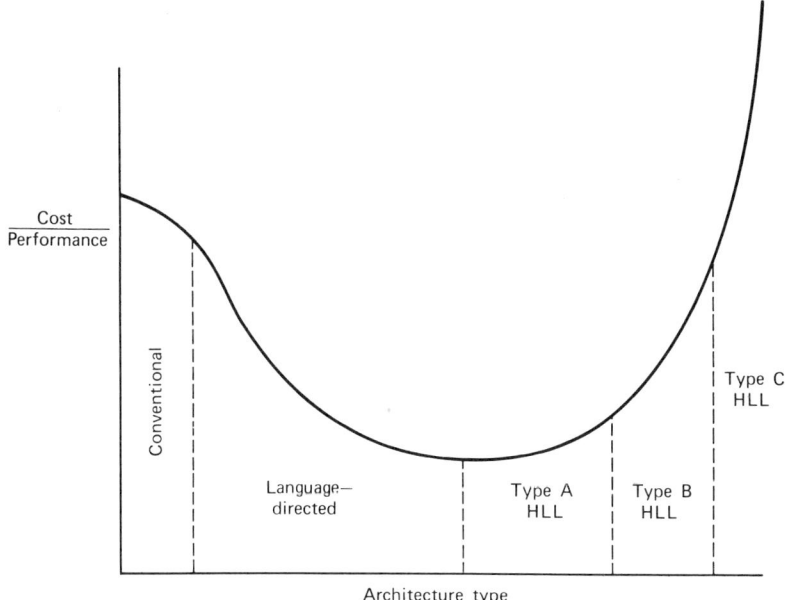

Figure 3.2 Relative cost/performance ratios among architectural approaches.

One other disadvantage of the type-C approach is that there is no function (i.e., compilation or assembly function) that examines the program prior to execution. One advantage of a compiler or assembler that is often overlooked is its detection of syntax errors. In a type-C machine, syntax errors would not be detected until execution time (e.g., a syntax error in a particular statement might not be detected until 30 minutes into the execution of the program). Of course, one could write a preprocessor to detect syntax errors, but then one might as well translate the source program into a more palatable form for machine execution, making the machine a type-A or type-B machine.

In summary, the type-C approach is not discussed further in this book because it fails to recognize that there are system functions that are best implemented in software and that there are transformations on source programs that are best performed prior to execution.

APPLICATION-DIRECTED ARCHITECTURES

The last category of architectures encompasses machines that are not necessarily oriented toward one or more programming languages, but toward specific applications. Since it is difficult to consider how to orient a general-purpose system toward specific end-use applications, most application-directed architectures are oriented toward one well-understood time-consuming application: the operating system.

As an example, the Honeywell Series 60 Level 64 architecture is oriented toward the efficient execution of operating-system functions [53,54]. For instance, the process-management functions (e.g., process creation, scheduling /dispatching, and synchronization) traditionally found in operating systems were instead placed within the machine. In recognition that operating systems spend a considerable amount of time managing queues and lists, special operating-system-oriented instructions were defined to search queues and lists and insert or delete entries in them.

Similarly, several models of the IBM S/370 series perform operating-system functions that are traditionally performed in software on other systems, although these functions were added after the fact rather than incorporated in the original architecture. In an experimental architecture within IBM [55], the machine performs such functions as program-to-program communication, process management, and storage management, functions normally associated with software operating systems. The IBM 3838 array-processing attachment to the S/370 allows the S/370 to continue functioning as a general-purpose system, but it also allows programs performing complex operations on large arrays to ship these operations to an auxiliary processor for more efficient computation.

If one wishes to orient an architecture toward end-use applications, the problems are more difficult, but several ideas have surfaced. One might consider a system with a dynamic instruction set and containing heuristics to optimize its architecture. One proposal [56] has the machine modifying its instruction set through such heuristics as looking for common sequences of instructions, generating a new instruction that can be used as an alternative to each sequence, and periodically informing the compilers of these changes. Another proposal [57] is to provide only a small kernel of machine instructions and allow the compilers to generate microprograms for new machine instructions (e.g., to replace commonly generated sequences of instructions or for the high-frequency sections of the program being compiled). It should be noted that, although such proposals are interesting, they are currently unproven.

Application-directed architecture is actually a second dimension of computer architecture, because such an architecture could also be a von Neumann, language-directed, or high-level-language architecture. In fact, many existing architectures in the last two categories are found to be application directed. For instance, the SYMBOL, B1700, and SWARD architectures discussed in later chapters embody many traditional operating-system functions; they could be considered to be application-directed (toward the operating system application) as well as language-directed or high-level language architectures.

REFERENCES

1. W. M. McKeeman, "Language Directed Computer Design," *Proceedings of the 1967 Fall Joint Computer Conference*. Washington, D.C.: Thompson, 1967, pp. 413–417.
2. L. N. McMahan and E. A. Feustel, "Implementation of a Tagged Architecture for Block Structured Languages," *Proceedings of the ACM-IEEE Symposium on High-Level Language Computer Architecture*. New York: ACM, 1973, pp. 91–100.
3. R. A. Brooker, "Influence of High-Level Languages on Computer Design," *Proceedings of the IEE*, 117(7), 1219–1224 (1970).
4. E. A. Hauck and B. A. Dent, "Burroughs B6500/7500 Stack Mechanism," *Proceedings of the 1968 Spring Joint Computer Conference*. Washington, D.C.: Thompson, 1968, pp. 245–251.
5. A. C. D. Haley, "The KDF.9 Computer System," *Proceedings of the 1962 Fall Joint Computer Conference*. Washington, D.C.: Spartan, 1962, pp. 108–120.
6. J. S. Miller and W. H. Vandever, "Instruction Architecture of an Aerospace Multiprocessor," *Proceedings of the ACM-IEEE Symposium on High-Level Language Computer Architecture*. New York: ACM, 1973, pp. 52–60.
7. M. J. Lutz and M. J. Manthey, "A Microprogrammed Implementation of a Block Structured Architecture," *Record of the Fifth Annual Workshop on Microprogramming*. New York: ACM, 1972, pp. 28–41.
8. T. Kilburn, D. Morris, J. S. Rohl, and F. H. Sumner, "A System Design Proposal," *Proceedings of IFIP Congress 1968*. Amsterdam: North-Holland, 1969, pp. 806–811.
9. C. H. Lindsey, "Making the Hardware Suit the Language," in J. E. L. Peck, Ed., *ALGOL 68 Implementation*. Amsterdam: North-Holland, 1971, pp. 347–365.
10. C. McFarland, "A Language-Oriented Computer Design," *Proceedings of the 1970 Fall Joint Computer Conference*. Montvale, N.J.: AFIPS, 1970, pp. 629–640.
11. B. W. Wade and V. B. Schneider, "A General-Purpose High-Level Language Machine for Minicomputers," *Proceedings of the ACM SIGPLAN-SIGMICRO Interface Meeting*. New York: ACM, 1973, pp. 169–171.
12. B. W. Wade and V. B. Schneider, "The L-Machine: A Computer Instruction Set for the Efficient Execution of High-Level Language Programs," *Record of the Fifth Annual Workshop on Microprogramming*. New York: ACM, 1972, pp. 81–82.
13. M. Sugimoto, "PL/I Reducer and Direct Processor," *Proceedings of the 24th ACM National Conference*. New York: ACM, 1969, pp. 519–538.
14. D. B. Wortman, "A Study of Language Directed Computer Design," Ph.D. dissertation, Stanford University, 1972.
15. R. J. Chevance, "A COBOL Machine," *Proceedings of the ACM SIGPLAN-SIGMICRO Interface Meeting*. New York: ACM, 1973, pp. 139–144.
16. S. M. Nissen and S. J. Wallach, "The All Applications Digital Computer,"

Proceedings of the ACM-IEEE Symposium on High-Level Computer Architecture. New York: ACM, 1973, pp. 43–51.

17. S. M. Nissen and S. J. Wallach, "An APL Microprogramming Structure," Record of the Sixth Annual Workshop on Microprogramming. New York: ACM, 1973, pp. 50–57.

18. W. K. Gilio and H. Berg, "STARLET—A Computer Concept Based on Ordered Sets as Primitive Data Types," Proceedings of the Second Annual Symposium on Computer Architecture. New York: IEEE, 1975, pp. 201–206.

19. M. C. Harrison, "A Language-Oriented Instruction Set for the BALM Language," Proceedings of the ACM SIGPLAN-SIGMICRO Interface Meeting. New York: ACM, 1973, pp. 161–168.

20. S. Tafuelin and A. Wikstrom, "Aspects of Compact Programs and Directly Executed Languages," BIT, **15**(2), 203–214 (1975).

21. M. D. Shapiro, "A SNOBOL Machine: A Higher-Level Language Processor in a Conventional Hardware Framework," Digest of the Sixth Annual IEEE Computer Society International Conference. New York: IEEE, 1972, pp. 41–44.

22. A. N. Myamlin and V. K. Smirnov, "Computer with Stack Memory," Proceedings of IFIP Congress 1968. Amsterdam: North-Holland, 1969, pp. 818–823.

23. E. A. Feustel, "The Rice Research Computer—A Tagged Architecture," Proceedings of the 1972 Spring Joint Computer Conference. Montvale, N.J.: AFIPS, 1972, pp. 369–377.

24. J. K. Iliffe, Basic Machine Principles. Second Edition. London: Macdonald, 1972.

25. H. C. Kancler, "Architecture of Aerospace Computer Simplifies Programming," Computer Design, **15**(5), 159–166 (1976).

26. A. Carpino, "An Architecture Development Tool for an Intermediate Language Machine (ILM)," Digest of the Fall Compcon75. New York: IEEE, 1975, pp. 265–267.

27. J. P. Dorocak, "Structured Control Operators Implemented on an Intermediate Language Machine (ILM)," Digest of the Fall Compcon75. New York: IEEE, 1975, pp. 268–271.

28. G. J. Myers, "The Design of Computer Architectures to Enhance Software Reliability," Ph.D. dissertation, Polytechnic Institute of New York, 1977.

29. J. K. Broadbent and G. F. Coulouris, "MEMBERS—A Microprogrammed Experimental Machine with a Basic Executive for Real-Time Systems," Proceedings of the ACM SIGPLAN-SIGMICRO Interface Meeting. New York: ACM, 1973, pp. 154–159.

30. W. C. Nielsen, "Design of an Aerospace Computer for Direct HOL Execution," Proceedings of the ACM-IEEE Symposium on High-Level Language Computer Architecture. New York: ACM, 1973, pp. 34–42.

31. A. J. Melbourne and J. M. Pugmire, "A Small Computer for the Direct Processing of FORTRAN Statements," The Computer Journal, **8**(1), 24–27 (1965).

32. M. F. de la Guardia and J. A. Field, "A High Level Language Oriented Multiprocessor," Proceedings of the 1976 International Conference on Parallel Processing. New York: IEEE, 1976, pp. 256–262.

33. R. M. Balzer, "An Overview of the ISPL Computer," Communications of the ACM, **16**(2), 117–122 (1973).

34. P. S. Abrams, "An APL Machine," Ph.D. dissertation, Stanford University, 1970.

35. A. Hassitt, J. W. Lageschulte, and L. E. Lyon, "Implementation of a High Level Language Machine," Communications of the ACM, **16**(4), 199–212 (1973).

REFERENCES

36. A. Hassitt and L. E. Lyon, "An APL Emulator on System /370," *IBM Systems Journal*, **15**(4), 358–378 (1976).
37. E. W. Reigel, U. Faber, and D. A. Fisher, "The Interpreter—A Microprogrammable Building Block System," *Proceedings of the 1972 Spring Joint Computer Conference*. Montvale, N.J.: AFIPS, 1972, pp. 705–723.
38. S. C. Schroeder and L. E. Vaugh, "A High Order Language Optimal Execution Processor," *Proceedings of the ACM-IEEE Symposium on High-Level Language Computer Architecture*. New York: ACM, 1973, pp. 109–116.
39. R. Zaks, D. Steingart, and J. Moore, "A Firmware APL Time-Sharing System," *Proceedings of the 1971 Spring Joint Computer Conference*. Montvale, N.J.: AFIPS, 1971, pp. 179–190.
40. B. J. Robinet, "Architectural Design of an APL Processor," in Y. Chu, Ed., *High-Level Language Computer Architecture*. New York: Academic, 1975, pp. 243–268.
41. T. R. Bashkow, A. Sasson, and A. Kronfeld, "System Design of a FORTRAN Machine," *IEEE Transactions on Electronic Computers*, **EC-16**(4), 485–499 (1967).
42. H. Weber, "A Microprogrammed Implementation of EULER on IBM System /360 Model 30," *Communications of the ACM*, **10**(9), 549–558 (1967).
43. M. Wells and A. Denson, "Direct Execution of Programming Languages," *The Computer Journal*, **17**(2), 130–134 (1974).
44. L. S. Haynes, "The Architecture of an ALGOL 60 Computer Implemented with Distributed Processing," *Proceedings of the Fourth Annual Symposium on Computer Architecture*. New York: IEEE, 1977, pp. 95–104.
45. See the SYMBOL references at the end of Chapter 8.
46. K. J. Thurber and J. W. Myna, "System Design of a Cellular APL Computer," *IEEE Transactions on Computers*, **C-19**(4), 291–303 (1970).
47. J. P. Anderson, "A Computer for Direct Execution of Algorithmic Languages," *Proceedings of the 1961 Eastern Joint Computer Conference*. New York: Macmillan, 1961, pp. 184–193.
48. H. M. Bloom, "Conceptual Design of a Direct High-Level Language Processor," in Y. Chu, Ed., *High-Level Language Computer Architecture*. New York: Academic, 1975, pp. 187–242.
49. J. E. Meggitt, "A Character Computer for High-Level Language Interpretation," *IBM Systems Journal*, **3**(1), 68–78 (1964).
50. J. C. Shaw, A. Newell, H. A. Simon, and T. O. Ellis, "A Command Structure for Complex Information Processing," *Proceedings of the 1958 Western Joint Computer Conference*. New York: AIEE, 1958, pp. 119–128.
51. A. P. Mullery, R. F. Schauer, and R. Rice, "ADAM—A Problem-Oriented Symbol Processor," *Proceedings of the 1963 Spring Joint Computer Conference*. Baltimore: Spartan, 1963, pp. 367–380.
52. J. Petit et al., "A Microprogrammed Strategy for HLL Interpretation," *SIGMICRO Newsletter*, **7**(4), 46–68 (1976).
53. T. D. Atkinson, "Architecture of Series 60 /Level 64," *Honeywell Computer Journal*, **8**(2), 94–106 (1974).
54. T. D. Atkinson, U. O. Gagliardi, G. Raviola, and H. S. Schwenk, Jr., "Modern Central Processor Architecture," *Proceedings of the IEEE*, **63**(6), 863–870 (1975).
55. G. Radin and P. R. Schneider, "An Architecture for an Extended Machine with Protected Addressing," TR-00.2757. Poughkeepsie, N.Y.: IBM Poughkeepsie Laboratory, 1976.

56. A. M. Abd-Alla and D. C. Karlgaard, "Heuristic Synthesis of Microprogrammed Computer Architecture," *IEEE Transactions on Computers,* **C-23**(8), 802–807 (1974).
57. T. G. Rauscher and A. K. Agrawala, "Developing Application Oriented Computer Architectures on General Purpose Microprogrammable Machines," *Proceedings of the 1976 National Computer Conference.* Montvale, N.J.: AFIPS, 1976, pp. 715–722.

EXERCISES

3.1 As implied in Exercise 2.1, an architecture that shrinks the semantic gap to one language might have an excessive semantic gap with respect to other languages. List several strategies that might overcome this problem.

3.2 Are the approaches in this chapter most applicable to large-scale or small-scale (e.g., minicomputer) systems?

3.3 Do you see anything unusual in the languages to which the referenced architectures are oriented?

4 | Requisites for Improved Architectures

Based on the problems in conventional architectures, a set of solutions can be developed to shrink the semantic gap. These solutions are not as trivial as simply increasing the "power" of a machine's instruction set (e.g., by adding a DO-loop-control instruction to a conventional architecture); rather, they involve radical changes to the memory concept of an architecture and represent departures from the classical von Neumann model. The solutions also involve increases in the power (function) of the machine instruction set, but this is a secondary fallout effect from the changes to the concept of memory.

The solutions are discussed only at a conceptual level in this chapter. The case-study architectures in the subsequent parts of the book are used to illustrate particular implementations of these ideas.

SELF-DEFINING DATA

The most significant step in closing the semantic gap, and the largest departure from the classical von Neumann model, is the idea of making all data words within memory self-identifying (also known as the concept of "tagged" or "typed" data). In the von Neumann architecture, it is the instruction that defines the attributes of the operands (e.g., the S/370 has distinct instructions for adding 31-bit binary integers, 15-bit binary integers, 32-bit floating-point numbers, 64-bit floating-point numbers, 128-bit floating point numbers, 1-digit decimal numbers, 3-digit decimal numbers, and so on). The alternative is to add a set of bits to each word in storage that describe the attributes of that word, as shown in Figure 4.1.

The self-identification field (or "tag") might identify the data type of the word (i.e., bits in the tag would denote whether the value field represents a binary integer, a decimal integer, a floating-point number, a character string, an address, etc.). One result is that the machine instructions can be defined as generic. That is, rather than having a large repertoire of add instructions, the machine need have only one; the type of addition to be performed is deduced by the machine by examining the tags of the operands of each instruction. The tag fields are transparent to the high-level-language programs; the tags are set by the compiler and are used by the machine to determine the type of operation to be performed, for consistency checks (e.g., the architecture might be defined such that the machine would refuse to add an address to a floating-point number), and for automatic data conversion (e.g., the architecture might be defined such that, if an add instruction refers to an integer operand and a floating-point operand, the addition is performed based on established data conversion rules).

In addition to identifying the type or representation of the word (Feustel suggests 32 possible types [1]), the concept of tags can be further exploited by the architect. For instance, if the data can be of varying length, a field could be defined in the tag to designate the length of the operand. Among other things, this would eliminate the repetition of the length information in each instruction that references each data word. For instance, the S/370 has 15 distinct add instructions. One of these, ADD DECIMAL, contains two 4-bit fields specifying the length of its two operands. In a sense, then, the S/370 has 270 distinct add instructions (14 plus 256 varieties of ADD DECIMAL). In a tagged architecture in which each tag described the type and length of the operand, there would be a need for only one add instruction.

Other possible extensions of the tag come to mind. One could add a 1-bit field describing whether the operand's value is currently defined, allowing the machine to detect attempts to use an undefined value. One could also add trap bits to the tag. For instance, a bit might be defined such that, whenever a program references this word, a machine interrupt is generated, providing a base for sophisticated debugging tools.

Figure 4.1   Self-identifying storage word.

SELF-DEFINING DATA

To illustrate the use of tags, Figure 4.2 represents the format of a fixed-point number in the SWARD architecture discussed in Part V. In this architecture, tags have different lengths and formats depending on what they are representing, but the first 4-bit field always defines the type of the storage area. In Figure 4.2, the 1110 in the first field designates this as a fixed-point number. The second field defines the number of digits of data. The third field (fraction size) describes the number of digits to the right of an imaginary decimal point. The first four bits of the data represent the sign, and the remaining bits (in binary coded decimal) represent the value. For instance, the number 5.74 would be stored as E320574 (in hexadecimal notation). Rather than defining a special tag bit to indicate an undefined value, "undefined" in this machine is represented as a particular data value. For fixed-point numbers, an undefined value would be indicated by four 1-bits at the start of the last field (e.g., E320F00).

The advantages of the concept of self-identifying or tagged data are

1. The machine can detect many classes of programming errors. For instance, the architecture can be defined such that errors are triggered if the operands of an instruction are not meaningful (e.g., an operand of a multiply instruction is a character string), incompatible (e.g., a program is attempting to store a floating-point number into an address), or if a source operand has an undefined value.
2. The machine can perform automatic data conversions if the operands are compatible but have different lengths or representations. For instance, the architecture can be defined such that it is legal to add an integer to a floating-point number, and the machine will perform a data conversion if it fetches an add instruction with these operands.
3. Because the machine performs the two processes just mentioned, the system will be faster than a conventional system that provides similar function. On a conventional system, run-time error checking and automatic data conversion must be accomplished through

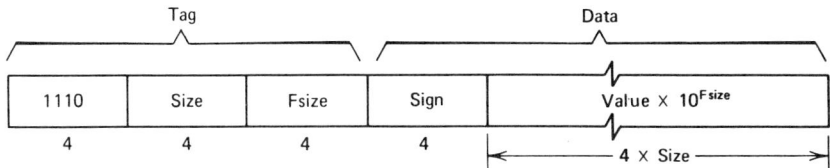

Figure 4.2 Fixed-point decimal cell in the SWARD architecture.

software (additional compiler-generated code); this implies additional processor/memory data transfers (for the additional instructions) plus additional instruction decodings.
4. The machine will have substantially fewer instruction types.
5. Compilers will be less complicated (and thus faster). In a conventional system, the code-generation process is complex because the compiler must do a semantic analysis of the program to determine the proper machine instructions to be generated. For instance, when a compiler encounters the + operator, it must examine the expressions on both sides of the operator to know which add instruction to generate. In a tagged machine, however, the compiler would simply generate the generic add instruction. Code generation is further simplified in that the machine, not the compiler, is responsible for run-time checks and automatic data conversion.
6. The construction of program debugging tools will be simplified. At the least, a crude debugging tool for a tagged architecture can produce a more meaningful storage dump (i.e., printing data in its proper representation) than a conventional dump of memory as a long sequence of binary, octal, or hexadecimal digits. Also, the possibility of placing trap bits in tags increases the feasibility of sophisticated debugging tools.
7. The concept of tagged storage aids the implementation of the data-independence concept in data base systems. Among other things, data independence implies that an application program can view a data base record in a way that is different from the physical representation of the record. (As a simple example, an application program can decide to view a field as a decimal number although in the data base the field might be represented in base two.) This concept is difficult to implement in current systems because machine instructions contain information about the type and lengths of their operands, making recompilation necessary whenever the data base definitions change. In a tagged architecture, however, if instructions are generic and tags are extended to secondary storage, implementation of data independence becomes more feasible.

Storage Requirements of Tags

The most common criticism of the tag concept is the feeling that tags increase the system's cost by increasing the amount of storage needed. For instance, intuition might tell us that adding a 4-bit tag to a system with 32-bit words increases the cost of storage by 12.5%. However, intuition is wrong here; in fact, there is reason to believe that a tagged-

SELF-DEFINING DATA

storage system might actually require fewer bits of storage than a conventional architecture.

The first reason is redundancy. If a floating-point operand is referenced by many instructions, each instruction redundantly identifies the operand as floating-point (because bits are needed in the operation-code field to distinguish floating-point operations from operations on other data types). Since variables tend to be referenced by more than one instruction, it makes more sense to store the static information (e.g., each operand's type) once with the operand rather than repeating it in each instruction.

As a simple example, consider an architecture X with 150 instruction types in which the operation code is expressed in an 8-bit field in the instruction. We wish to consider an alternative architecture Y using tags in which a three-bit tag is added to each data word in storage. Architecture Y can now be defined with generic instructions; a reasonable approximation to make is that Y needs only 50 instruction types, which can be represented in a 6-bit op-code field.

To compare the storage requirements of the two architectures, we can compare the number of op-code and tag bits (call this sum B) needed in a program, making the assumption that all other storage requirements are constant. If I represents the number of machine instructions in the program, then B for machine X is $8I$. On machine Y, B is equal to $6I$ plus three times the number of operands in the program. If each instruction in both machines references two operands, and if each operand in the program is referenced an average of R times, then B on machine $Y = 6I + 6I/R$. The ratio of B for machine Y over B for machine X is $0.75 \times (1 + 1/R)$; if this ratio is less than 1, the tagged machine requires less storage.

Evaluating the ratio requires knowledge of R, the average number of references to each operand. One would expect R to be greater than 1. The cutover point is 3; at this point the two machines require the same amount of storage. This number R is not a widely measured factor, but one measurement of R for a set of programs yielded the value 10.4 [2]. If this value is representative, one would expect the space needed for op-codes and tags in a typical program on this tagged architecture to be 82% of the space needed for op-codes on the untagged architecture.

Hence a tagged architecture might require fewer storage bits because of the elimination of redundancy in instruction op-codes. This counter-intuitive phenomenon substantiates a point in Chapter 1: the architect must have a thorough understanding of programming languages and the characteristics of programs to make intelligent decisions.

A tagged architecture also reduces a program's storage requirements by eliminating a second type of redundancy: the instructions that are repeatedly generated by compilers to perform run-time checks and automatic data conversion. As an example, assume that PL/I variable I is declared as FIXED BINARY (31) and A and D are declared as FIXED DECIMAL (4, 2). Using IBM's PL/I Optimizing Compiler on the S/370, the generated code for the statement D=D+A occupies 6 bytes of storage, but the code for D=D+I occupies 64 bytes (an increase of 967%), as shown in Table 4.1. The difference is due to the code generated by the compiler for the data conversion. If one enables the option to check for an overflow when storing into D, the statement D=D+I generates 92 bytes of object code. As indicated in Table 4.1, the increase in execution time is even more significant; executing the mixed-mode statement D=D+I takes a whopping 3011% more time than the statement D=D+A (measured on a S/370 Model 145).

This conversion and checking code is typically generated in-line where needed. (It need not be; the code could exist in one place and subroutine calls could be generated, but this usually proves to be excessively inefficient.) A tagged architecture eliminates this redundancy by delegating these functions to the machine. Of course, the functions must exist somewhere; for instance, in a microprogrammed machine, they occupy space in control storage. However, in a tagged machine the functions exist only once (in control storage) rather than being replicated hundreds or thousands of times in the programs residing in main storage.

The third reason why tags do not require as much storage as first appears is that they need not be added to every word. For instance, there is no need to store a tag with every element in an array. Since, by definition, array elements have identical attributes, the tags can be factored out; only one tag is necessary for the array. Similarly, there is no need to tag each element of a character string; only one tag is necessary to define the attributes of the elements.

Table 4.1 The Cost of Software Automatic Data Conversion and Overflow Checking

Statement	Machine Instructions	Size (bytes)	Execution Time (microseconds)
D=D+A;	1	6	13.1
D=D+I;	13	64	407.6
(SIZE): D=D+A;	7	38	57.7
(SIZE): D=D+I;	19	92	448.1

SELF-DEFINING DATA OBJECTS

The self-identification concept can be carried even further to include less primitive and multidimensional data objects such as arrays and structures. Although tags could be used to describe such objects, a different identification method is often necessary because the objects are multidimensional. The form of identification is commonly called a *descriptor*. A descriptor is similar to a tag in that it describes the attributes of data, but it differs in that descriptors are disjoint from the data; the descriptor points to the data, and instructions indirectly address the data through the descriptors.

Descriptors can take many forms, but as a short example we examine the use of descriptors to define arrays in the Burroughs B6700. The word size in the B6700 is 51 bits; the first three bits represent a tag. If a tag has the value 101, the word is a descriptor. One type of descriptor, a word descriptor, is illustrated in Figure 4.3. The P (presence) bit indicates whether the described data is in main or secondary storage. The C (copy) bit indicates whether this descriptor is a copy of another descriptor. The I (indexed) bit indicates whether the descriptor points to an entire aggregate of data, or just one element within the aggregate. The S (segmented) bit indicates whether the described data is contiguous or segmented in memory. The R (read-only) bit is set if only reads are permitted to the described data. The T bits, which would be zero in this case, define this descriptor to be a word descriptor (as opposed to a string descriptor). The D bit indicates whether the described data is single or double precision. If the I bit is off, the next field in the descriptor defines the number of elements of described data, and the last field points to the beginning of the described data.

The B6700 descriptors can be combined into tree structures to describe multidimensional storage objects. For instance, Figure 4.4 shows the storage structure for a three-by-four array of words. The array descriptor points to a three-element vector of descriptors which in turn point to four-element vectors of data words. One implication is that the machine is responsible for array indexing calculations and bounds checking. To obtain a particular element, the machine-language pro-

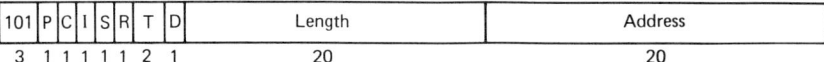

Figure 4.3 B6700 word descriptor.

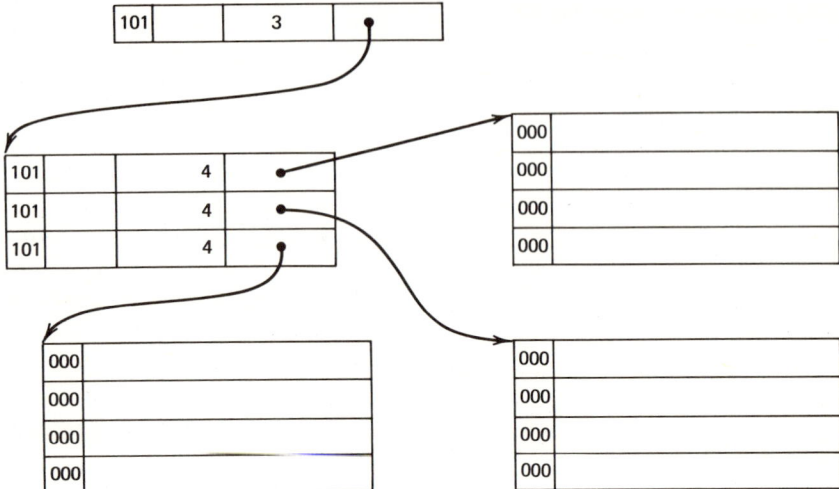

Figure 4.4   Representation of a two-dimensional array.

gram simply names the array descriptor and the values of the subscripts. The machine calculates the address of the referenced element and, at the same time, determines whether the subscripts are within the defined bounds.

The advantages of self-defining data objects such as arrays and structures are

1. Because the machine is now able to perform such operations as array indexing, the system is faster than a conventional system (largely due to the decrease in memory accesses for instructions and the decrease in instruction decodings).
2. The concept of generic instructions can be extended to these data objects (e.g., if the ADD instruction points to two array operands, the machine can be defined to do an element-by-element addition). Again this is a performance advantage, because the machine is not wasting time by repeatedly fetching and decoding software instructions that accomplish an element-by-element addition (refer to the example in Chapter 2).
3. Because these advantages imply fewer generated machine instructions, the storage requirements of programs are reduced. In Chapter 2 we saw that the PL/I statement

    C(I,J) = A(I,J) + B(J,I);

SELF-DEFINING DATA OBJECTS

generates 17 S/370 instructions occupying 62 bytes. Most of these instructions are generated for the array-indexing operations, and these instructions are redundantly repeated in storage for every array reference in the program.

4. For the reasons just mentioned, the code-generation processes of compilers are simplified.
5. Making the machine aware of multidimensional data objects, accompanied by the use of generic instructions, opens up new possibilities for the processor designer. One obvious possibility is the ability to perform certain operations in parallel. A more important advantage is the flexibility given to the processor designer faced with the development of a compatible family of machines. After one exploits existing circuit technology to the fullest and widens a processor's data paths, current processor designers are rather limited in their opportunities to create faster "top-of-the-line" processors. A higher-level architecture, however, opens up new avenues. The low-end processors could be purely sequential in operation, but the high-end processors might exploit the opportunities for parallelism inherent in instructions manipulating multidimensional data objects. Also, processors with cache memories are usually limited to loading the cache on a demand basis, but if the processor is aware of such objects as arrays and structures, it can prefetch data into the cache more intelligently.
6. The machine is now able to detect such programming errors as the use of an array subscript whose value is beyond the bounds of the corresponding array dimension. The overhead of such a check by the machine is negligible, particularly in light of the performance gains provided by descriptors. In Chapter 2 we saw why most compilers find a software check for this error impractical; if the optional SUBSCRIPTRANGE check is enabled for the PL/I statement

    C(I,J) = A(I,J) + B(J,I);

    57 S/370 instructions are executed rather than 17, and the size of the generated code is 274 bytes rather than 62 bytes.

Although the concepts of tags and descriptors have been described separately, the concepts are quite similar, and an appropriate generalization of the architecture could merge the two concepts. For instance, the architecture in Part V generalizes the concept of tags to allow tags to describe multidimensional data objects as well as scalar data.

EXPRESSION-EVALUATION STACKS

A third architectural advance that has been claimed to shrink the semantic gap between computer architectures and languages is the use of push-down stacks for the evaluation of expressions (e.g., arithmetic expressions in the form of arithmetic statements and logical expressions in IF and DO statements). Rather than beginning with a discussion of the concepts of a stack-oriented instruction set, it is helpful to examine an example.

Assume that a compiler is examining the arithmetic statement

M = A*B+C/(D+E)

This statement can be easily transformed into a parentheses-free form called the postfix or reverse Polish form with the appearance

MAB*CDE+/+=

If the underlying machine has a stack-oriented instruction set, the instruction stream can be generated directly from this form by generating one instruction per symbol. To see how this is done, assume the underlying machine contains the instructions

1. PUSH—this pushes the value of the specified operand onto the top of the stack.
2. PUSHAD—this pushes the address of the specified operand onto the top of the stack.
3. STORE—this places the top value of the stack into the storage word pointed to by the second element on the stack and then deletes the top two elements on the stack.
4. ADD, MULT, DIVIDE—these instructions perform an arithmetic operation on the top two elements of the stack and then remove the top two elements and push the result onto the top of the stack.

This instruction set is compared to a conventional register-oriented instruction set. Assume the latter machine has both register-to-register and register-to-storage instructions and that the op-codes in both machines occupy 8 bits and storage addresses in the instructions occupy 20 bits (we do not concern ourselves with how these 20 bits are defined). Assume also that a register address in the latter machine's instruction set occupies 4 bits. Assuming that the operands of the statement M=A*B+C/(D+E) are one-word integers, the instruction streams generated for the two machines are

EXPRESSION-EVALUATION STACKS

PUSHAD	M	LOAD	R1,A
PUSH	A	MULT	R1,B
PUSH	B	LOAD	R2,C
MULT		LOAD	R3,D
PUSH	C	ADD	R3,E
PUSH	D	DIVREG	R2, R3
PUSH	E	ADDREG	R1, R2
ADD		STORE	R1, M
DIVIDE			
ADD			
STORE			

The stack machine receives six 28-bit instructions and five 8-bit instructions; the register machine receives six 32-bit instructions and two 16-bit instructions. Although three more instructions were generated for the stack machine than the register machine, the advantages of the stack-oriented instruction set appear to be

1. The stack-oriented instruction stream occupies 208 bits, but the register-oriented instruction stream occupies 224 bits. The difference is due to the use of the stack as an implied address (e.g., the target of the PUSH instruction is always the stack, but the target of the LOAD instruction must be explicitly identified; both operands of the stack-oriented arithmetic instructions are implied and thus occupy no space in the instruction stream). Thus, in addition to stack-oriented instruction sets appearing to result in smaller storage requirements, fewer bits must pass between the processor and main storage, a performance advantage.
2. The number of units of information that must be examined and interpreted by the instruction-fetch portion of the processors is less for the stack-oriented instruction set, another performance advantage. Counting op-codes, register addresses, and storage addresses as units, the stack machine must examine 17 units versus 24 for the register machine in the previous example.
3. The code-generation process of a compiler is simpler for a stack-oriented instruction set. In addition, the compiler for a register machine must manage the use of the registers. (True, a stack machine is more complicated in that the processor must manage the stack, but this complexity exists in only one place, rather than existing in each compiler.)

Are Evaluation Stacks Important?

Although the use of stacks for expression evaluation appears to be superior to the use of registers, the differences are not very significant, and certainly the advantages of stacks do not appear to be as significant as the advantages of tagged storage and data descriptors.

Ignoring the differences between stack-oriented and register-oriented instruction sets for the moment, both have been claimed to have two advantages over another alternative: storage-to-storage addressing. The main advantage is storage economy. For instance, 16 or 20 bits are needed in a S/370 instruction to address an operand in main storage, but only 4 bits are needed if the operand is in a register. The second advantage is performance; the number of instruction bits transferred is less, simpler address calculations are needed when an operand is in a register or on the top of the stack, and the registers and the stack (or the top few positions on the stack) are usually represented in a faster storage technology than that of main storage. However, registers and stacks also have several disadvantages:

1. Registers and stacks are oriented toward fixed-size operands. If the architecture has no fixed-size-word concept, the use of registers and stacks is cumbersome.
2. If the architecture is designed such that direct storage addresses in instructions are relatively short (the machine in Part V has such an architecture), the major advantage of stacks and registers is less significant.
3. Although registers and stacks might be contained in faster storage devices, one must weigh this against the time and space wasted by instructions that move operands between main storage and the registers or stack. Such "overhead" instructions are unnecessary if all addressing is storage-to-storage.
4. The advantages of evaluation stacks are only significant when the compiled program makes use of them. A favorite example in compiler textbooks is showing how a complex arithmetic or logical expression (e.g., the earlier example) can be represented elegantly in reverse Polish notation, but there is considerable doubt that such situations arise very often in actual programs. One study [3] disclosed that 21,078 expressions contained just 15,969 operators, or an average of 0.76 operators per expression. Another study of 440 FORTRAN programs [4] produced similar results: 60% of the assignment statements contained no operators other than the assignment ("=") operator. Data from 120 commercial PL/I programs [5] shows that 98% of all expressions have zero or one operator.

Points 3 and 4 are verified by a study of the frequency distribution of instructions in the Hewlett-Packard HP3000 system [6]. The measured programs included data base inquiry programs, a BASIC interpreter, a text editor, several compilers, and a sort. Of the instructions executed, only 16% were zero-address stack-to-stack instructions. Furthermore, 25% of the instructions executed were push and pop (load and store) instructions.

The author feels that stack-oriented instruction sets are not an important advance in computer architecture, and that storage-to-storage addressing, in actual rather than contrived situations, is more efficient and more closely models the concepts in high-level languages (see Exercises 4.7–4.10). One reason for introducing the concept of evaluation stacks, however, is that the first few case-study architectures in this book have stack-oriented instruction sets.

SUBROUTINE MANAGEMENT

A major way of closing the semantic gap is to make the machine aware of the concepts of program structure, particularly the important concept of the subroutine or procedure. The motivation for this is the lengthy process performed by software on conventional machines just to create the subroutine concept. For instance, a PL/I program must perform these steps when a procedure is called: (1) a block of storage called an activation record is dynamically allocated to hold the local variables and status information of the called procedure; (2) this activation record is added to the software-maintained stack of activation records for the program; (3) the status of the calling procedure is saved in its activation record; (4) local storage and the parameters of the called procedure are initialized; and (5) a branch is taken to the called procedure.

The alternative is to provide the appropriate call/return and storage-management mechanisms in the machine architecture. To illustrate this, assume that storage is organized as multiple pushdown stacks (called activation stacks), one per program. (This use of stacks is not necessarily associated with the evaluation stack discussed in the previous section and does not imply a stack-oriented instruction set.) An entry in an activation stack is a variable-size activation record, which contains the storage for a subroutine's local variables and parameters, an instruction counter, a pointer to the previous activation record in the stack, and possibly some status information.

The machine might have call and return instructions that are semantically similar to the CALL and RETURN/END statements in programming languages. When a call instruction is executed, the machine builds a new activation record and adds it to the activation stack (which implies that each subroutine or procedure probably has associated descriptive information defining its attributes, such as its parameters and its local storage). When a return instruction is executed, the machine removes the top activation record from the stack.

The advantages of such a subroutine-management architecture should be obvious: increased performance and decreased program size. Also, if the tagged-data concept is present, the machine can efficiently check for such errors as inconsistent arguments and parameters.

LEXICAL-LEVEL ADDRESSING

The use of stacks for subroutine management has ramifications on the manner in which storage is addressed. For instance, an operand in an activation record has no fixed address, since the activation record may appear at different locations within the stack during the program's execution. A second motivation for a different addressing mechanism is that there is no need (or desire) for an instruction to be capable of addressing any location in storage. In fact, an instruction within a subroutine or procedure must only be capable of addressing the data defined within the subroutine (local variables, parameters, and any global data to which it has access). A third motivation is closing the semantic gap by providing an addressing mechanism that more closely resembles the concepts of addressing in high-level languages.

One addressing concept that satisfies these needs is the concept of lexical-level addressing, which is oriented toward block-structured languages (e.g., PL/I and ALGOL). A lexical-level address is represented as an address couple consisting of two items: the lexical level within the program of the variable or operand and the index (sequential number) of the operand within that lexical level. Since the lexical ordering of the program and the indexes of operands remain static during a program's execution (although the physical locations of operands may not), the lexical-level address can be used as an operand address in machine instructions. The machine can maintain the appropriate information to dynamically translate lexical-level addresses into physical addresses.

This concept is best illustrated by an example; the PL/I program in Figure 4.5 is used. The outer procedure resides at lexical level 0. Four identifiers are defined in level 0: A, B, VVV, and WWW (A and B are variables and VVV and WWW are procedure names; thus VVV and WWW might represent descriptors to the respective procedures). As shown in Figure 4.5, the lexical-level addresses of these identifiers are (0,1)–(0,4).

LEXICAL-LEVEL ADDRESSING

```
                                    ADDRESS    REFERENCE
UUU: PROCEDURE;
    DECLARE A;                      (0,1)
    DECLARE B;                      (0,2)
    CALL VVV;                                  (0,3)
    CALL WWW;                                  (0,4)
    VVV: PROCEDURE;                 (0,3)
        DECLARE C;                  (1,1)
        CALL XXX;                              (1,2)
        XXX: PROCEDURE RECURSIVE;   (1,2)
            DECLARE D;              (2,1)
            C=8;                               (1,1)
            A=D;                               (0,1),(2,1)
            CALL XXX;                          (1,2)
        END;
    END;
    WWW: PROCEDURE;                 (0,4)
        DECLARE E;                  (1,1)
        E=8;                                   (1,1)
        B=8;                                   (0,2)
    END;
END;
```

Figure 4.5 Lexical-level-address assignments and references.

There are two procedures at lexical-level 1; thus the identifiers within these procedures have the address couples (1,x) as shown. This illustrates an interesting property of lexical-level addressing: addresses are not unique. For instance, variables C and E have the same address; thus the machine instructions generated for the statements C=8 and E=8 would be identical (e.g., symbolically the machine instructions might be MOVE,1,1,'8'). However, this property of nonunique addresses presents no problem; as we see shortly, the machine can unambiguously resolve the address (1,1) to the proper physical location.

To see how lexical-level addresses are resolved, assume that the machine is currently executing procedure WWW. (For now, forget about the possibility of recursive calls such as the one in XXX, because this requires a slightly different mechanism for address translation.) There will be two activation records on the activation stack, the top one for WWW and the next one for UUU, and the machine can easily "remember" that it is executing in lexical level 1 (assume that this number is stored in a register named R). When the machine encounters an address (X,Y), it might compute the formula

Physical address = physical address of activation
                    record (R−X) from top of stack + Y

For instance, when the machine executes the code for statement E=8, it encounters the address (1,1), which is translated to the address of the first operand in activation record 0 (the top one) on the stack. When it encounters the address (0,2) (for B), it translates this to the second operand in activation record 1 (the activation record for UUU).

Unfortunately, this simple address-translation process is not sufficient, for it does not work properly in the event of recursive invocations of procedures. When recursive calls occur, the order of activation records does not directly correspond to lexical levels. For instance, if procedure XXX has called itself twice, there are five records on the activation stack—three for XXX and one each for VVV and UUU—and the simple formula just given would produce incorrect addresses. Therefore the machine needs a more dynamic method of keeping track of the correspondence between lexical levels and activation records. This method involves the use of an array of addresses usually referred to as *display registers* [7].

To illustrate the use of display registers, assume that the previous PL/I program has executed the sequence of statements

CALL VVV;/CALL XXX;/C=8;/A=D;/CALL XXX;

The content of the activation stack is shown in Figure 4.6. As records are added to and removed from the activation stack, the machine updates the set of display registers so that each display register points to the activation record corresponding to each lexical level. The machine resolves the address (1,1) for the machine instruction(s) corresponding to the statement C=8 by adding 1 to the value of display register 1. To resolve the address (0,1), it adds 1 to the value of display register 0. The machine's call-return mechanism is responsible for maintaining the display registers.

Although lexical-level addressing is considered here as an architectural advance, its future applications are unclear. One reason is that it is oriented toward block-structured languages. but it represents a form of "overkill" for such languages as COBOL, FORTRAN, and RPG. Second, its primary purpose is the resolution of scope-of-name rules (i.e., if some variable is used but not declared in an inner block, it is globally shared with some other block), but there are people ([8–11] to cite a few) who suggest that this is an undesirable programming practice.

CAPABILITY-BASED ADDRESSING

A second addressing concept that appears to be more promising because of its increased generality is known as capability-based or unique-name addressing [12–14]. In an architecture with this form of

CAPABILITY-BASED ADDRESSING

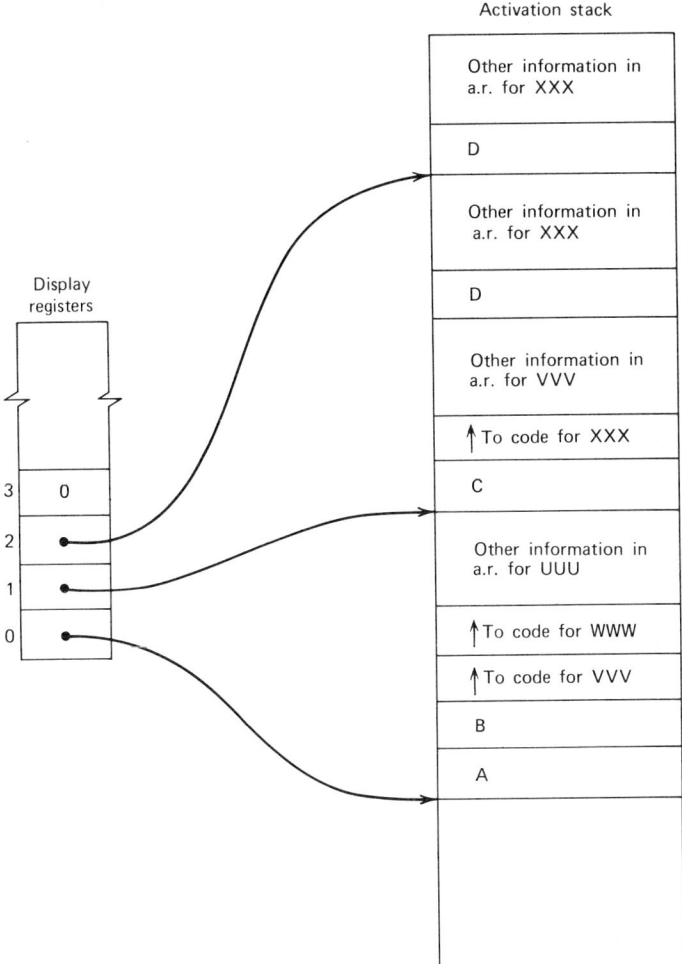

Figure 4.6 Use of display registers to locate activation records (a.r. = activation record).

addressing, storage is viewed as a *set* of named objects. Programs refer to storage through the object names, but have no idea where a storage object resides in storage.

A *capability* may be thought of as a systemwide address of a collection of data. A capability usually contains three components:

1. A unique system-wide name of a storage object. Storage objects might represent such entities as activation records, program modules, and dynamically allocated data segments, and, depending on

the architecture, this concept might also be extended to name such entities as processes, files, and I/O devices.
2. A displacement or index of a datum within the storage object.
3. An access or authority code controlling the type of access (e.g., read/write, read-only) permitted to the storage object.

When an object is created, the machine assigns it a unique name and gives the program that created the object a capability to it. A program may create copies of capabilities or give a capability to another program, but programs are prohibited from fabricating capabilities on their own or modifying the name within a capability. (Note that this implies the use of tagged storage, in which "capability" is a distinct data type.)

Every machine instruction that addresses storage must implicitly or explicitly specify a capability for that storage. To a program, a capability is simply a set of bits that uniquely identify a piece of data, but, to the machine, a capability is a bit pattern that must must be transformed to yield a physical address. Hence the machine must maintain the correspondence between names in capabilities and actual storage locations, which implies the use of hash tables, associative registers, or both.

The name in a capability is chosen by the machine when an object is created. The assigned name is unique in that no two objects have the same name; once an object is deleted, its name is "never" reassigned to another object. Although capability-based addressing has been used without this property of unique names, the property is considered vital because of its advantages. If the names are unique and programs are prohibited by the machine from fabricating names on their own and from modifying names, the concept provides a powerful storage-protection mechanism, since the only storage that can be referenced by a program is storage that the program creates and storage to which the program is given a capability by another program. The unique-name concept also assists the machine in detecting certain types of software errors, particularly those associated with the use of pointer or reference variables in such languages as PL/I. This use of capabilities is discussed in the architecture in Part V.

Since a capability must be represented in a finite number of bits, the architect must approximate the phrase "never reassigned" as "not reassigned for a sufficently long time." For instance, if the architect estimates that a unique name will be assigned once per 100 microseconds on the average and if a reasonable approximation of "never" is 2 years, a 40-bit name is sufficient. (The simplest way of assigning names is for the machine to maintain a permanent counter and assign names sequentially. When the counter reaches the maximum value, it recycles to the value zero. The counter must maintain its value when power is removed from the machine.)

Capability-based addressing does not imply that machine instructions contain one or more large capabilities to address their operands. In fact, capabilities cannot be imbedded in machine instructions because, at compilation time, the capabilities would not have been assigned. With some creative design, a capability-based architecture can have relatively short operands in instructions. For instance, the architecture in Part V uses capability-based addressing, yet the length of typical machine instructions is from 16 to 28 bits.

Both capability-based and lexical-level addressing imply that programs cannot arbitrarily create addresses, thus providing a storage-protection mechanism. Both addressing mechanisms also imply that storage does not have the appearance of a linear sequential von Neumann memory and that the machine is responsible for storage management on a logical level (e.g., allocating storage objects) and a physical level (e.g., moving or paging information through a hierarchy of storage devices).

VARIABLE-SIZE STORAGE CELLS

Another way of reducing the semantic gap is eliminating the concept of fixed-size storage words at the architecture interface. As we saw in Chapter 2, fixed-size words lead to storage wastage because of the nonuniform distribution of data values, to language distortions, and to execution inefficiencies and further storage consumption because of the code that must be generated to map the programmer's data representations into fixed-size words.

There is not much to say about variable-size storage cells in a conceptual sense other than that a method of representation is to describe the size of each cell within its tag (see Figure 4.2). As is the case for other information in tags, if there is uniformity in the sizes of a collection of cells (e.g., elements in an array), the size information can be factored out and represented only once.

REFERENCES

1. E. A. Feustel. "On the Advantages of Tagged Architecture," *IEEE Transactions on Computers*," **C-22**(7), 644–656 (1973).
2. E.C.R. Hehner, "Computer Design to Minimize Memory Requirements," *Computer*, 9(8), 65–70 (1976).
3. W. G. Alexander and D. B. Wortman, "Static and Dynamic Characteristics of XPL Programs," *Computer*, 8(11), 41–46 (1975).

4. D. E. Knuth, "An Empirical Study of FORTRAN Programs," *Software—Practice and Experience*, **1**(2), 105–133 (1971).
5. J. L. Elshoff, "An Analysis of Some Commercial PL/I Programs," *IEEE Transactions on Software Engineering*, **SE-2**(2), 113–120 (1976).
6. R. P. Blake, "Exploring a Stack Architecture," *Computer*, **10**(5), 30–39 (1977).
7. E. A. Hauck and B. A. Dent, "Burroughs B6500/B7500 Stack Mechanism," *Proceedings of the 1968 Spring Joint Computer Conference*. Washington, D.C.: Thompson, 1968, pp. 245–251.
8. W. Wulf and M. Shaw, "Global Variable Considered Harmful," *SIGPLAN Notices*, **8**(2), 28–34 (1973).
9. G. J. Myers, *Reliable Software Through Composite Design*. New York: Petrocelli/Charter, 1975.
10. J. B. Dennis, "Modularity," in M. Beckman, G. Goos, and H. P. Kunzi, Eds., *Advanced Course in Software Engineering*. Berlin: Springer-Verlag, 1973, pp. 128–182.
11. G. Goos, "Language Characteristics: Programming Languages as a Tool in Writing System Software," in M. Beckman, G. Goos, and H. P. Kunzi, Eds., *Advanced Course in Software Engineering*. Berlin: Spring-Verlag, 1973, pp. 47–69.
12. J. B. Dennis and E. C. Van Horn, "Programming Semantics for Multiprogrammed Computations," *Communications of the ACM*, **9**(3), 143–155 (1966).
13. R. S. Fabry, "Capability-Based Addressing," *Communications of the ACM*, **17**(7), 403–412 (1974).
14. T. A. Linden, "Operating System Structures to Support Security and Reliable Software," *Computing Surveys*, **8**(4), 409–445 (1976).

EXERCISES

4.1 What language construct might be troublesome to implement in an architecture with tagged storage?

4.2 To convince yourself that the tag concept reduces, rather than increases, storage requirements, compute the R-factor (the average number of references in machine instructions to an operand) for an existing program. Since the generated machine instructions, not the source statements, must be analyzed, use a compiler that can produce an assembly-language listing. (IBM's PL/I Optimizing Compiler has such an option.)

4.3 Similar to the earlier example on the storage requirements of tags, consider a machine X with an 8-bit op-code and two 4-bit fields in each instruction that define the lengths of the operands. (The IBM S/370 has this appearance when doing decimal arithmetic or character-string handling, i.e., when executing a COBOL program.) Consider a machine Y with generic instructions requiring a 6-bit op-code and having an 8-bit tag associated with each data item in storage (the tag defines the data items' type and length).

EXERCISES

Assuming that each instruction references two operands and that there are an average of R references per operand, at what value of R does machine Y require less storage?

4.4 In exercise 4.3, assume that R is 10, operand addresses occupy 16 bits, and the average operand size is 32 bits (excluding the tag). How much space does a program occupy on Machine Y compared to the same program on machine X?

4.5 If one was designing an architecture with self-defining arrays and the system's only programming language was FORTRAN, would it be sensible to extend the concept of generic instructions to allow entire arrays as operands?

4.6 Using the instructions defined in the section on evaluation stacks, what machine instructions would be generated for the statement A=B+B*B?

4.7 Since program data show that the majority of expressions have zero or one operator, consider the statement A=A+B. How many instructions would be generated for this statement in evaluation-stack, general-register, and two-address storage-to-storage architectures?

4.8 Counting op-codes and addresses as units, how many units of information must be interpreted by the machine for each of the three approaches?

4.9 Assuming 8-bit op-codes, 4-bit register addresses, and 20-bit storage addresses, what is the size of the instruction stream for each of the three approaches?

4.10 In addition to the statement A=A+B, the most common forms of the assignment statement are A=B and A=B+C. Considering this, what can one deduce about the relative value of the three approaches?

II
A Language-Directed Architecture

5 | The Student-PL Machine

As an initial case study to illustrate many of the ideas in Part I, the architecture of the Student-PL Machine (SPLM) [1] is examined. The original purpose of SPLM was to serve as an example of how one would optimize or tune a language-directed architecture; this has several ramifications on our study of the machine. First, the machine has not been physically implemented (i.e., it exists only on paper). Second, the architecture is incomplete (e.g., functions that would be needed by an operating system are missing from the architecture). Last, we examine the initial unoptimized version of the architecture, because optimization is treated separately in Chapter 17 (the optimization of SPLM is used as an example, however, in Chapter 17). These ramifications are not serious obstacles, because SPLM is a simple useful illustration of many of the concepts in Part I.

SPLM is a language-directed architecture, directed toward the Student-PL language, a subset dialect of PL/I (i.e., a subset of PL/I that is not fully compatible with PL/I). In relation to the concepts in Chapter 4, SPLM contains tagged storage, descriptors of data objects, pushdown stacks for expression evaluation, subroutine management, and PL/I-controlled storage, and lexical-level addressing. However, contrary to the ideas in Chapter 4, SPLM uses fixed-size storage words and contains only binary arithmetic.

THE STUDENT-PL LANGUAGE

Since SPLM is directed toward the Student-PL language, the appropriate starting point is a brief informal discussion of the language. The statements, data types, built-in functions, and operators in the language are listed in Table 5.1.

The FIXED data type is equivalent to FIXED BINARY (31,0) in PL/I, FLOAT is equivalent to FLOAT BINARY (21), BIT is equivalent to BIT (1), and CHARACTER is equivalent to CHARACTER (4095) VARYING in PL/I. All variables except arrays are in the automatic storage class. All arrays are in the PL/I controlled storage class, meaning that array storage must be explicitly allocated via the ALLOCATE and FREE statements.

The Student-PL statements are closely related to their PL/I counterparts. A program may be partitioned into begin blocks and internal procedures and functions. The GO TO statement is more restricted than in PL/I; its target cannot be within a compound statement structure (e.g., within a THEN or ELSE clause or within a DO loop). Rather liberal

Table 5.1 Student-PL Language

Statements	Data Types	Built-in Functions	Operators
PROCEDURE	FIXED	ABS	+
BEGIN	FLOAT	ATAN	−
RETURN	BIT	BIT	*
END	CHARACTER	CHARACTER	/
DO	LABEL	COS	**
DO WHILE	ENTRY	DUMP	<
DO CASE	Scalar	E-FORMAT	>
Iterative DO	Array	END-OF-DATA	=
CALL		EXP	<=
GO TO		FIXED	>=
DECLARE		FLOAT	¬<
Assignment		INDEX	¬>
IF		LENGTH	¬=
ALLOCATE		LOG	||
FREE		MAX	
		MIN	
		MOD	
		RANDOM	
		SIGN	
		SIN	
		SQRT	
		SUBSTR	
		TAN	
		TIME	
		UNDEFINED	

automatic data conversion rules exist for the evaluation of arithmetic, logical, and string expressions, as in PL/I. PL/I array arithmetic is present in Student-PL

Student-PL also contains 25 built-in functions which are similar to their PL/I counterparts. As in PL/I, some of them may be used as pseudovariables (i.e., may appear as the target of an assignment statement). The built-in function INPUT and the pseudovariable OUTPUT are semantically equivalent to the PL/I GET LIST and PUT LIST statements. The BIT, CHARACTER, FIXED, and FLOAT functions are used to force a data conversion; INDEX, LENGTH, and SUBSTR are string-related functions, and UNDEFINED is used to test a variable for an undefined value.

SPLM STORAGE STRUCTURE

The best place to begin the study of an architecture is at its concept of storage. The compiler writer or machine-language programmer views SPLM as having seven distinct memories, four registers, and an array of 16 registers, as shown in Figure 5.1. Each memory and register has a distinct purpose, as indicated in Figure 5.1. The memories and registers are logically related in that their values often point to locations within other memories; these intermemory relationships are indicated by the arrows in Figure 5.1.

Although SPLM has no concept of a "main memory," the principal memory of interest is the data stack memory. The bottom portion of this memory is used as a pushdown stack for two purposes: to serve as a stack for expression evaluation and to hold the local storage for all active procedures, functions, and begin blocks within the program. The DSP register always points to the top word in this stack. The top portion of the data stack memory is managed by the machine; it is used for the storage of array elements.

The data stack memory is a tagged memory. Its word size is 39 bits, which is subdivided into a 7-bit tag and 32-bit value, as shown in Figure 5.2. If the u-bit in a word's tag is set, the word has the "undefined" value. If the p-bit is set, the word represents a procedure (or function) pointer. If the a-bit is set, the word represents an array pointer. The remaining 4 bits in the tag contain type information. The meaning of the 32-bit value field is dependent on the word's tag, as indicated in Table 5.2. In some cases this field is interpreted as two distinct fields of 12 and 20 bits. In such cases the meaning of the two fields is listed in parentheses.

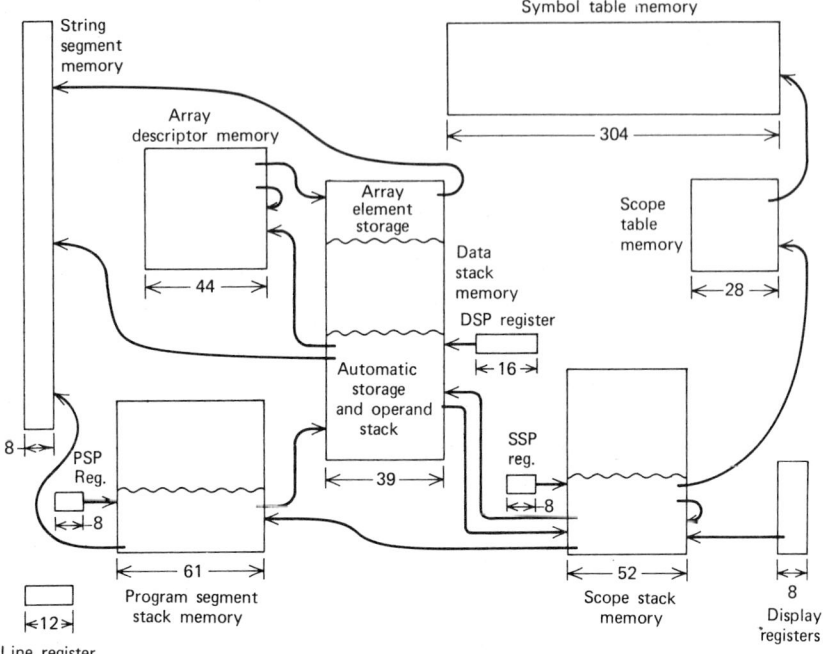

Figure 5.1 SPLM storage structures.

To give a simplified explanation of some of these words, arrays are represented as array pointers, character strings as character-string descriptors, addresses of variables as indirect-address descriptors, addresses of array elements as array-element descriptors, and addresses of object-program segments as program-segment descriptors. The "star" word is used to represent a subscript with the * value, as in the PL/I expression A(*,I).

The string segment memory has a word size of 8 bits and contains object programs and character strings. Object programs are stored as one or more segments. A segment represents a procedure, function, begin block, DO-loop body, or THEN or ELSE clause of an IF statement.

The array descriptor memory contains descriptors of arrays. An entry is inserted by the machine whenever an array is dynamically allocated. Each entry is represented in D+1 contiguous words, where D is the number of dimensions in the array. An array descriptor for a two-dimensional array is shown in Figure 5.2. The first field contains the address of the first array element in the data stack memory. The subsequent words contain information about each dimension (i.e., the defined lower and upper subscript bounds and a multiplier used to

Data stack memory word

1 1 1	4	12	20
u p a	Type		

Array descriptor memory entry for a two-dimensional array

16	16	12
Ptr. to first element in data stack	No. of dimensions	Ptr. to array descriptor for previous allocation
Lower bound	Upper bound	Multiplier
Lower bound	Upper bound	Multiplier

Program segment stack memory word

12	20	13	16
Length of segment	Ptr. to first instruction in string segment memory	Offset of next inst. to be executed	Value of DSP at time of first inst.

Symbol table memory word

256	8	8	32
Symbol name	No. of dimensions	Type	Initial value

Scope table memory word

8	8	8	4
Ptr. to first symbol table entry	No. of symbol table entries	No. of parameters	Lexical level

Scope stack memory word

8	8	16	8	12
Ptr. to scope table entry	Ptr. to SSM entry for prev. lexical level	Value of DSP at time of entry	Value of PSP at time of entry	Value of line register at time of entry

Figure 5.2 SPLM storage-word formats (ptr. = pointer).

Table 5.2 Data Stack Memory Words

Tag	Meaning of the Value Field
u-bit=1	Undefined value
p-bit=1	Program-segment descriptor (segment length, address of segment in string segment memory)
a-bit=1	Array pointer (unused, address of entry in array descriptor memory)
type=fixed	Signed binary integer
type=float	Signed binary floating-point value
type=bit	Boolean value (true or false)
type=char	Character-string descriptor (string length, address of string in string segment memory)
type=labc	Label constant descriptor (offset in program segment, address of program segment in string segment memory)
type=labv	Label variable descriptor (offset in program segment, address of program segment in string segment memory)
type=star	None (represents a * subscript)
type=ind	Indirect-address descriptor (address of scope stack entry, offset from DSP)
type=arlm	Array-element descriptor (address of entry in array descriptor memory, offset of element in array)

calculate the span between elements in that dimension). The third field in the first word is used to implement the concept of PL/I controlled storage (e.g., if array A is allocated and then allocated again, there are two array descriptors for A, and the current descriptor points, via this field, to the previous descriptor).

The program segment stack is used to maintain the history of active program segments. A source program is partitioned by the compiler into one or more program segments. Each procedure, function, begin block, THEN clause, ELSE clause, and DO-loop body is represented as a separate program segment. The reason for representing THEN and ELSE clauses and DO loops as separate segments is explained in the next chapter; it is also shown that this is a weakness in the architecture.

An entry is placed on the program stack whenever a segment is entered; the top entry is deleted whenever a segment completes execution. The PSP register always points to the top entry. One can view the top entries on the program segment stack and the scope stack as the machine's "PSW" (in S/370 terms), since they define the current state of the machine. The four fields in the program segment stack word in Figure 5.2 are self-explanatory. The last field contains the value that was in the DSP register at the time of executing the first instruction in the segment.

The symbol table memory is initialized by the compiler or program loader and is used by the machine when allocating local storage for a segment. The symbol table contains one entry per identifier and constant in the source program. The format of an entry is shown in Figure 5.2. The 8-bit type field is used to initialize a tag in the data stack memory when space for the variable is allocated.

The scope table memory is also initialized by the compiler or program loader. The memory contains one entry per change in lexical level in the program (i.e., one entry per procedure, function, and begin block). As shown in Figure 5.2, a scope table entry describes the symbol table entries associated with the level and defines the number of parameters to be received and the lexical level.

The last memory is the scope stack, which is used to maintain the history of active procedures, functions, and begin blocks. An entry is placed on the scope stack whenever a procedure, function, or begin block is entered. As shown in Figure 5.2, the scope stack entry points to the corresponding scope table entry and the scope stack entry for the previous lexical level and contains the values of the DSP, PSP, and line registers at time of entry.

The machine also contains an array of 16 display registers, which are used as described in Chapter 4 to resolve lexical-level addresses. The display registers are numbered from 0 to 15; display register n contains the address of the scope stack entry corresponding to the active lexical-level n of the program.

The last register, the line register, has a passive role. It is used to keep track of which source statement is currently executing; thus if an error occurs, the error can be reported to the programmer in terms of his or her source program. When the compiler is generating code corresponding to source statement 17, for example, the first generated machine instruction is LINE 17, which loads the value 17 into the line register.

REFERENCE

1. D.B. Wortman, "A Study of Language Directed Computer Design," Ph.D. dissertation, Stanford University, 1972. Also available as report CSRG-20, University of Toronto, 1972.

EXERCISES

5.1 In a data stack word with the a-bit set (i.e., the word is an array pointer), what is the purpose of the 4-bit type field?
5.2 In what memories does the compiler or program loader place its output?
5.3 What memories are not altered during the program's execution?
5.4 How would an array of character strings be represented?
5.5 What is the maximum number of lexical levels in a Student-PL program?
5.6 From the descriptions of the memories and registers, determine which are altered when a Student-PL CALL statement is executed.
5.7 Why does a scope stack entry contain an explicit pointer to the entry for the next lowest lexical level? That is, wouldn't this entry always be the prior one on the stack?

6 | Program Compilation and Execution on SPLM

Rather than studying the SPLM architecture instruction by instruction, a more meaningful way to study an architecture is by examining the compilation and execution of one or more high-level-language program examples. Hence this chapter analyzes SPLM by examining the compilation and execution of two Student-PL programs. Chapter 7 contains a traditional specification of the instruction set to which the reader may wish to refer during the discussions in this chapter.

SPLM has a stack-oriented instruction set. The first 8 bits of an instruction are the operation-code field. Most instructions consist of only an op-code and therefore have a length of 8 bits. A few instructions contain a second 8-bit field whose meaning is dependent on the instruction type. One instruction (LNAME) is a 24-bit instruction.

The first example is the compilation and execution of the simple procedure in Figure 6.1. Assume that this procedure is part of a larger program and that the procedure is at lexical level 2. The procedure contains two variables, A and B; their lexical-level addresses are (2,1) and (2,2). The procedure also contains the constants 1 and 3; these are assigned the addresses (2,3) and (2,4).

The compiler produces three results when compiling this procedure: an object-program segment in the string segment memory, four entries in the symbol table memory, and an entry in the scope table memory. The latter two results are shown in Figure 6.1; the object program is determined below (assume it is to be stored starting at location 446).

Although we are about to analyze the object code that would be

```
110    X: PROCEDURE;
120    DECLARE A FIXED;
130    DECLARE B FLOAT;
140    B = 1;
150    A = B + 3 * B;
160    END;
```

Symbol table memory entries

10	A		0	FIXED	
11	B		0	FLOAT	
12			0	FIXED	1
13			0	FIXED	3

Scope table memory entry

8	10	4	0	2

Figure 6.1   First SPLM example.

generated for this procedure, it is helpful to consider how the procedure is executed. Assume that this procedure was called by another procedure, which means that the last instruction executed in the other procedure is an ENTER instruction. The ENTER instruction allocates procedure X's local storage on the data stack, places entries on the program segment stack and scope stack, and updates the display registers. The resultant state of the machine might be that shown in Figure 6.2.

The next step is to consider the object code needed to represent procedure X. For ease of representation, the object code is represented in a hypothetical assembly language.

The first instruction in every procedure must be a SCOPEID instruction; its operand must be the address of the procedure's scope table entry. When SCOPEID is executed, it has absolutely no effect (i.e., it acts like a "no-op"). However, during the execution of the previous ENTER instruction, the machine references the SCOPEID instruction to locate the scope stack entry, which in turn defines the procedure's lexical level and points to the symbol table entries. Hence the first generated instruction is

SCOPEID   8

The next step is to generate the instructions for statement 140. The first instruction would be

LINE   140

which loads the statement number into the line register. To generate the remaining instructions, we consider the reverse Polish form of the statement, which is B1=. This indicates the sequence of instructions to be generated, namely, instructions that push the address of B onto the

PROGRAM COMPILATION AND EXECUTION ON SPLM

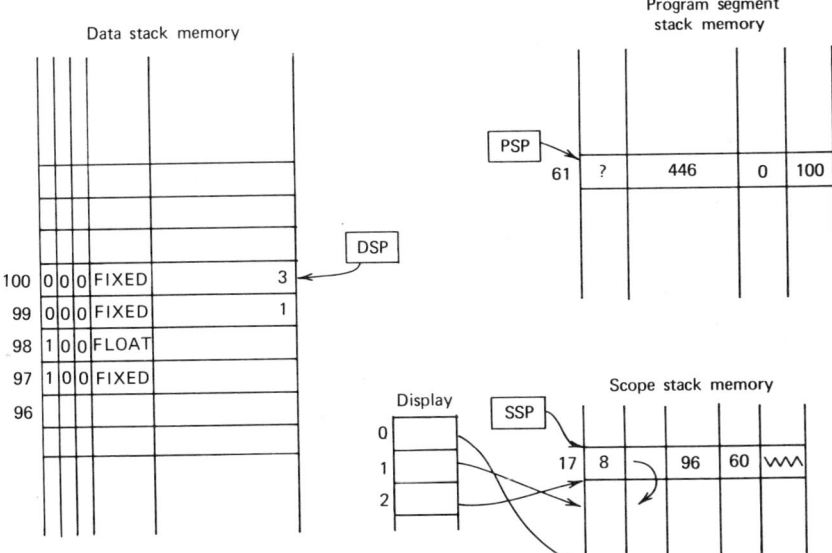

Figure 6.2 Machine state on entry to procedure X.

stack, push the value 1 onto the stack, and perform a store. The first instruction is

SNAME 2,2

The operand of this instruction is a lexical-level address (of B). Its function is to create an indirect-address descriptor (Table 5.1) and push it onto the stack. An indirect-address descriptor is a partially resolved lexical-level address. Its 32-bit value consists of a 12-bit address of a scope stack entry and a 20-bit offset from a prior value of the DSP register. The creation of an indirect-address descriptor is quite simple. Given the instruction SNAME x,y, the machine creates an indirect-address descriptor whose first 12 bits have the value of display register x and the remaining 20 bits have the value y. Therefore the instruction creates an indirect-address descriptor at data stack location 101 containing the value 17,2.

The next step is to get the value 1 on the stack. This is accomplished by the instructions

SNAME 2,3
EVAL

The first instruction creates an indirect-address descriptor with the value 17,3 at data stack location 102 (the DSP register now points to word 102). The EVAL instruction replaces the indirect-address descriptor on the top of the stack with the word referenced by the descriptor. To do this the machine goes to scope stack word 17, takes the DSP value stored in this word (96), adds three to it (99), and copies the data stack word at this address (the constant 1) onto the top of the stack. Hence data stack word 102 now contains a copy of word 99.

The next instruction generated is

STORE

This instruction deals with the top two stack words. It expects the second word to be a descriptor (in this case it is an indirect-address descriptor to B). It copies the top word in the stack into the word pointed to by the descriptor, and then deletes the second word. Hence word 98 (B) now contains the value 1, and its undefined bit is turned off. Note that a FIXED value is being stored into a FLOAT word; the machine recognizes this and performs the appropriate data conversion to floating point.

The statement B=0 is now complete, but DSP has the value 101 (the value 1 is still on the stack); the stack is cleaned up by the instruction

POP

which deletes the top word from the stack (i.e., subtracts 1 from DSP).

The code for statement 150 is generated in a similar way. The reverse Polish representation of the statement is AB3B*+=; thus the generated code is

```
LINE       150      ...
SNAME      2,1      ind(A), ...
SNAME      2,2      ind(B),ind(A), ...
EVAL                1,ind(A), ...
SNAME      2,4      ind(3),1,ind(A), ...
EVAL                3,1,ind(A), ...
SNAME      2,2      ind(B),3,1,ind(A), ...
EVAL                1,3,1,ind(A), ...
MUL                 3,1,ind(A), ...
ADD                 4,ind(A), ...
STORE               4, ...
POP                 ...
```

PROGRAM SEGMENTS FOR IF STATEMENTS AND DO LOOPS

The resultant contents of data stack words 101 and above are shown next to each instruction. The EVAL instruction also tests the operand for a defined value. For instance, if B had the undefined value, the first EVAL instruction would have generated a program interrupt. The arithmetic instructions (MUL and ADD) perform their operations on the top two words and then delete these two words and push the result onto the stack. If their operands have different data types (which is the case for the MUL instruction), an automatic data conversion is performed during instruction execution.

The remainder of the generated code is

```
LINE   160
END    8
```

The END instruction returns control to the calling procedure by deleting the local storage on the data stack, deleting the top entries on the program segment stack and scope stack and updating the display registers. In this case the END instruction would simply set the DSP register to 96, the PSP register to 60, the SSP register to 16, and display register 2 to zero.

PROGRAM SEGMENTS FOR IF STATEMENTS AND DO LOOPS

One unusual property of SPLM is that it has no branch instruction. It does have a GOTO instruction, but this instruction is intended to correspond to a Student-PL GO TO statement; the GOTO instruction branches to the location contained in a label-constant or label-variable descriptor.

The absence of a conventional branch instruction raises the question of how one would generate code for IF statements and DO loops. The architect's reasoning was that an IFTHENELSE statement can be considered a conditional subroutine call. If the IF condition is true, the code representing the THEN alternative is called; if the IF condition is false, the code representing the ELSE alternative is called. In either case, control is returned to the statement following the IF statement. Likewise, a DO loop can be viewed as a call to the loop body followed by a return to the statement following the DO loop. The loop body can be viewed as an iteration of tests of the DO conditions followed by a call to the code within the loop.

This mechanism is best understood by example. Assume that this statement is added to procedure X in Figure 6.1:

```
155   IF   (A=1)   THEN B=3;
                   ELSE A=3;
```

Rather than generating one program segment, the compiler would now generate three segments. The two new segments are

```
SNAME      2,2              SNAME      2,1
SNAME      2,4              SNAME      2,4
EVAL                        EVAL
STORE                       STORE
POP                         POP
```

The first step in executing statement 155 is evaluating the IF condition; the generated code would begin as

```
LINE       155
SNAME      2,1
EVAL
SNAME      2,3
EVAL
EQ
```

The EQ instruction compares the top two stack words for equal values (doing any necessary data conversion) and replaces them with a bit variable on the top of the stack. Now, if we had previously set up a two-element one-dimensional array of program-segment descriptors of these two segments, the SUBS (subscript) instruction could be generated to use the bit value to select one of the program-segment descriptors, and then the CALL instruction is used to invoke the proper segment. (The name "CALL" was an unfortunate choice, because the CALL instruction simply places an entry on the program segment stack and transfers control to the segment. It is not associated with arguments or a change in lexical level. The instruction corresponding to the Student-PL CALL statement is the ENTER instruction.)

For statement 155, these instructions would be generated after the EQ instruction:

```
SNAME    2,5     ind of array of program-segment desc.
SWAP             exchange top two stack words
SUBS     1       puts an array-element descriptor on
                 the top of the stack (DSP now has the
                 value 101)
```

PROGRAM SEGMENTS FOR IF STATEMENTS AND DO LOOPS

EVAL replaces the above with the array
element (a program-segment desc.)
CALL branches to the program segment.
When the end of the segment is reached,
control returns here.

Unfortunately, this does not quite complete the compiler's output. An entry is needed in the symbol table for the array. Since arrays are not automatically allocated, ALLOCATE and FREE instructions must be added to allocate and free the array. Since the array must be initialized, two more entries (program-segment descriptors) are needed in the symbol table to initialize the array, and instructions to initialize the array must be generated. The complete compiler output for procedure X with the added IF statement is shown in Figure 6.3.

To understand the cost of this mechanism, assume that we add two new instructions to SPLM: BRANCH and BRANCH-FALSE. Both are 16-bit instructions; the 8 bits following the op-code is a signed value to be added to the instruction counter. For the BRANCH-FALSE instruction, the value is added to the instruction counter only if the bit value on the top of the stack is false. BRANCH-FALSE also removes the top word from the stack. With this change, the object program in Figure 6.3 can be rewritten as

SCOPEID 8		SNAME 2,2		SNAME 2,2	
LINE 140		EVAL		SNAME 2,4	
SNAME 2,2		MUL		EVAL	
SNAME 2,3		ADD		STORE	
EVAL		STORE		POP	
STORE		POP		BRANCH 9	
POP		LINE 155		SNAME 2,1	
LINE 150		SNAME 2,1		SNAME 2,4	
SNAME 2,1		EVAL		EVAL	
SNAME 2,2		SNAME 2,3		STORE	
EVAL		EVAL		POP	
SNAME 2,4		EQ		LINE 160	
EVAL		BRANCHF 11		END 8	

This minor change reduces the number of generated instructions to 39 (from 67) and has an even healthier effect on execution time, since the more time-consuming SUBS, ALLOC, FREE, and CALL instructions are eliminated.

PROGRAM COMPILATION AND EXECUTION ON SPLM

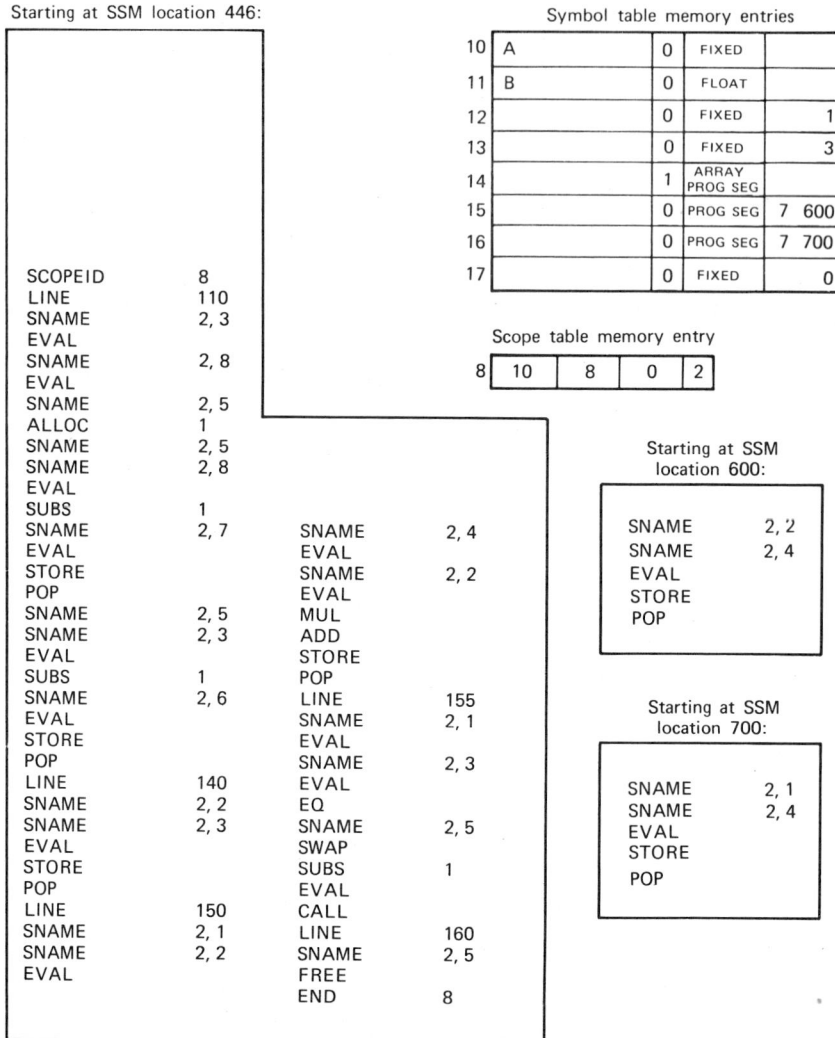

Figure 6.3 Compiler output for modified procedure X.

The lack of conventional branch instructions makes the implementation of DO-WHILE and iterative-DO loops even more complicated, too much so to be illustrated here. DO loops require two levels of segment calls, although some valuable instructions do exist (CYCLE, DOTEST, DOINCR; see Chapter 7) to implement DO loops.

This criticism of the IF and DO mechanisms in SPLM is not meant as an overall criticism of the machine, for the machine is being used as a

SUBROUTINE-CALL EXAMPLE

case study of some advances over conventional architectures. The criticism does point out, however, that the intent of an architecture should be to accurately and efficiently execute programs written in high-level languages. Beyond this, the architect should not believe, for instance, that the lack of a branch statement in the language or the desirability of "GO-TO-less" programming necessitates removing branch instructions from the architecture.

SUBROUTINE-CALL EXAMPLE

Since a substantial part of this architecture is concerned with automatic subroutine management and array operations, it is worthwhile to examine an example that introduces some of these concepts. This Student-PL program is used as the example:

```
 1 XYZ:   PROCEDURE;
 2 DECLARE A(*) FLOAT;
 3 DECLARE (B,C) FIXED;
 4 B=4;
 5 ALLOCATE A(1:B);
 6 A=1;
 7 C=B+B*B;
 8 CALL ZZZ(B);
 9 A(B)=C;
10     ZZZ:  PROCEDURE(M);
11     DECLARE M FIXED;
12     C=M+M;
13     END;
14 END;
```

Figure 6.4 illustrates the two program segments generated by the compiler. The machine instructions are numbered for ease of reference. Assume that the first segment begins at location 200 and the second at location 800 in the string segment memory. Figure 6.5 contains the remainder of the compiler's output, and Figure 6.6 shows the state of the machine when XYZ is entered.

Instructions 1–8 are straightforward and similar to those of the previous example. Instructions 9–14 allocate array A. The ALLOC instruction requires an indirect-address descriptor on the top of the stack for an array pointer, followed by the values of the lower and upper bounds. Note that source statement 6 performs an array operation: all elements of A are set to 1. Instructions 16–20 accomplish this. When the STORE

```
        XYZ Segment                    ZZZ Segment
  1  SCOPEID 1    27  SNAME 0,2    50  SCOPEID 2
  2  LINE 4       28  EVAL         51  LINE 12
  3  SNAME 0,2    29  MUL          52  SNAME 0,3
  4  SNAME 0,4    30  ADD          53  SNAME 1,1
  5  EVAL         31  STORE        54  EVAL
  6  STORE        32  POP          55  SNAME 1,1
  7  POP          33  LINE 8       56  EVAL
  8  LINE 5       34  SNAME 0,2    57  ADD
  9  SNAME 0,2    35  SNAME 0,6    58  STORE
 10  EVAL         36  ENTER 1      59  POP
 11  SNAME 0,5    37  POP          60  LINE 13
 12  EVAL         38  POP          61  UNDEF
 13  SNAME 0,1    39  LINE 9       62  END 2
 14  ALLOC 1      40  SNAME 0,1
 15  LINE 6       41  SNAME 0,2
 16  SNAME 0,1    42  EVAL
 17  SNAME 0,5    43  SUBS 1
 18  EVAL         44  SNAME 0,3
 19  STORE        45  EVAL
 20  POP          46  STORE
 21  LINE 7       47  POP
 22  SNAME 0,3    48  LINE 14
 23  SNAME 0,2    49  HALT
 24  EVAL
 25  SNAME 0,2
 26  EVAL
```

Figure 6.4 Object code for procedures XYZ and ZZZ.

instruction is executed, it recognizes that a scalar is being stored into an array; thus it stores the scalar into each array element.

Instructions 34–38 are generated as a result of the CALL statement. The operand of the ENTER instruction specifies the number of arguments. Descriptors for the argument and the program-segment descriptor must reside on the stack. The ENTER instruction verifies that the number of arguments is equal to the number of parameters, removes the

Symbol table memory entries

1	A	1	FLOAT		
2	B	0	FIXED		
3	C	0	FIXED		
4		0	FIXED	4	
5		0	FIXED	1	
6	ZZZ	0	PROG SEG	20	800
7	M	0	FIXED		

Scope table memory entries

1	1	6	0	0
2	7	1	1	1

Figure 6.5 Symbol table and scope table for procedures XYZ and ZZZ.

SUBROUTINE-CALL EXAMPLE

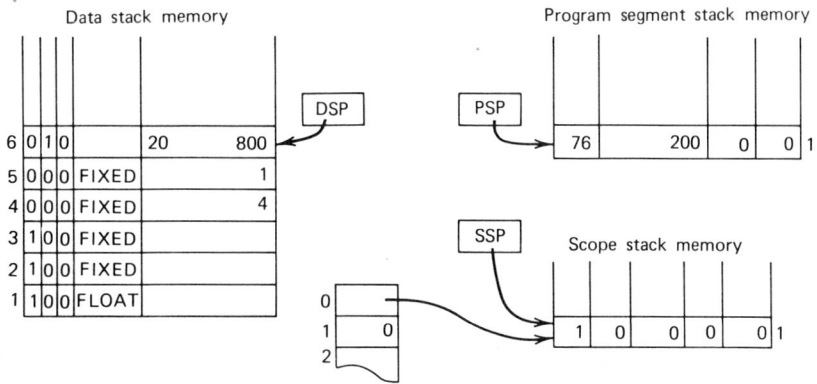

Figure 6.6 Machine state on entry to procedure XYZ.

pointer to the program-segment descriptor from the stack, allocates storage on the data stack for the called procedure (one word in this case for M), initializes the parameters as indirect-address descriptors to the arguments, places entries on the program segment and scope stacks, and updates the display registers. This causes procedure ZZZ to begin execution.

When instruction 54 is executed, the indirect-address descriptor on the top of the stack (as a result of 53) points to another indirect-address descriptor (the word representing M in word 8 on the data stack, which has been initialized as a parameter) which in turn refers to B. This presents no problem because the EVAL instruction is defined to follow a chain of descriptors. It brings the value at the end of the chain to the top of the stack.

The END instruction performs almost the opposite function of the ENTER instruction. However, since the END instruction is also used for returning from functions (which have a value), it expects the top data stack item to be a returned value. After END removes the local storage for this procedure from the stack, it places the returned value on the top of the stack. Since a procedure call has no value, the UNDEF instruction was generated to place an undefined word on the stack, and instruction 37 removes it.

Instructions 40–47 correspond to the statement A(B)=C. The effect of instructions 40–43 is to place an array-element descriptor on the top of the stack. If the value of B exceeds the bounds of the array, the SUBS instruction would generate a program interrupt. The store instruction stores the value of C (automatically converted to floating-point) into the element referenced by the array-element descriptor.

SIGNIFICANCE OF SPLM

To obtain a feeling for the merits of the SPLM machine versus a conventional machine the Student-PL program of the preceding section can be compared to an equivalent PL/I program compiled to the S/370. The equivalent PL/I program is

```
(SUBSCRIPTRANGE): XYZ: PROCEDURE;
DECLARE A(*) FLOAT BINARY(21) CTL;
DECLARE (B,C) FIXED BINARY(31);
B=4;
ALLOCATE A(1:B);
A=1;
C=B+B*B;
CALL ZZZ(B);
A(B)=C;
   ZZZ: PROCEDURE(M);
   DECLARE M FIXED BINARY(31);
   C=M+M;
   END;
END;
```

On SPLM the Student-PL program occupies 96 bytes of storage and executes 62 instructions. Considering the symbol table and scope table, the total program size is 369 bytes. The PL/I S/370 program's generated code occupies 564 bytes of storage, and the number of executed instructions is 303. The S/370 program occupies a total of 5912 bytes of storage because of the necessary run-time subroutines.

The comparison of 62 versus 303 executed instructions is not accurate; the S/370 number is actually much greater than 303, but a direct comparison is not meaningful. The PL/I Optimizing Compiler has a highly efficient call/return mechanism, but the penalty is a large number of instructions executed once at the beginning of the program to set up storage for the call/return mechanism. These instructions, a one-time cost per program, are excluded from the comparison.

One could argue, however, that the comparison is biased against PL/I because all Student-PL arrays must be dynamically allocated; however, it is not necessary to use PL/I controlled storage and the ALLOCATE statement for array A in the program. If the ALLOCATE statement is removed, A is removed from controlled storage, and SUBSCRIPTRANGE is disabled, the S/370 generated code occupies 360 bytes of storage, the number of executed instructions is 112, and the total program size is 5496 bytes.

In relation to the ideas of Part I, the significance of SPLM is that it uses tagged storage, many storage-management functions that are traditionally performed by operating systems or compiler execution-time subroutines are instead performed by the machine, programs cannot manipulate or fabricate addresses, and array operations, automatic data conversion, and subroutine linkage are performed by the machine. The machine also detects certain classes of programming errors, such as the use of an invalid subscript and references to undefined values. The data in the previous example shows that a language-directed architecture results in considerably smaller object programs than in a conventional machine. A language-directed machine can also be expected to significantly outperform a conventional machine because of significant reductions in the number of bits that must be transmitted between the processor and memory to execute a given high-level-language program and because of reductions in the number of machine cycles devoted to decoding machine instructions.

To be objective, SPLM also exhibits several problems. Its method of implementing IF statements and DO loops is cumbersome and inefficient. There is unnecessary redundancy in storage (e.g., every element in an array has an identical 7-bit tag, except possibly for the undefined bit). SPLM is a binary, rather than decimal, machine, and it uses fixed-size words. Because SPLM is an experimental paper machine, it is incomplete in many ways (e.g., operating-system and input/output considerations are absent). Because of its close association with the Student-PL language, compiling other languages to the machine would be difficult, or more likely, impossible (e.g., one could not write a PL/I compiler because it is impossible to implement certain PL/I constructs, such as structures and based variables, on the machine). However, the intent is not to criticize SPLM, but to use it as an initial case study of the implementation of many of the ideas in Part 1.

EXERCISES

6.1 In the first program segment in Figure 6.3, what is on the data stack prior to the execution of the ALLOC instruction?

6.2 In the first program segment in Figure 6.3, what is the purpose of the eight instructions following the ALLOC instruction?

6.3 How does the ENTER instruction locate the scope table entry of the segment being entered?

6.4 How does the ENTER instruction determine if the number of arguments is equal to the number of parameters?

6.5 In Figure 6.4, what is the purpose of the two POP instructions (instructions 37 and 38?)

6.6 When the ENTER instruction is initializing parameters, how does it determine which items in the symbol table represent parameters?

6.7 The best way to get a feel for an architecture is to perform a mental compilation of a source program. Invent a small Student-PL program, and compile it to SPLM. (All of the SPLM instructions are specified in Chapter 7.)

7 | SPLM Instruction Set

This chapter serves as a specification of the SPLM instruction set. The specification, in prose form, is derived from the original specification expressed in a machine description language. The actions taken after SPLM detects an error are omitted here, as they were omitted from the original specification.

For ease of understanding, the instructions are grouped into five categories. Instructions in the data-access and addressing category are used to fetch and store addresses and values to and from the data stack. The data-operation instructions perform arithmetic, logical, and string operations on their operands. The control instructions are related to program sequencing and the implementation of IF and DO statements. The procedure instructions are related to the management of procedures, functions, and begin blocks. The last category, array-storage instructions, contains the instructions that are related to storage allocation for arrays.

The instructions are described in terms of their effect on the data stack. The contents of the stack prior to and after the execution of each instruction are illustrated. The stack is illustrated horizontally; the leftmost symbol represents the top of the stack, and the symbol X represents lower items in the stack that do not play a part in the execution of the instruction.

Whenever an instruction is listed as using an indirect-address descriptor on the stack, that stack word may be an indirect-address descriptor to the eventual operand (one level of indirection), an indirect-address descriptor to another indirect-address descriptor, and so on, eventually pointing to an operand (multiple levels of indirection), or the operand itself.

DATA-ACCESS AND ADDRESSING INSTRUCTIONS

Instruction: SNAME (Short Name)
Instruction Size: 16 bits
Function: Push an indirect-address descriptor of the referenced word onto the stack.
Instruction Field: A lexical-level address. The first 4 bits represent a lexical-level, and the last 4 bits represent an index number.
Initial Stack: X
Final Stack: VAR,X
Operands: VAR is an indirect-address descriptor formed from the lexical-level address. If the referenced word is an indirect-address descriptor, its value is copied into VAR.

Instruction: LNAME (Long Name)
Instruction Size: 24 bits
Function: Push an indirect-address descriptor of the referenced word onto the stack.
Instruction Field: A lexical-level address. The first 4 bits represent a lexical-level, and the last 12 bits represent an index number.
Initial Stack: X
Final Stack: VAR,X
Operands: VAR is an indirect-address descriptor formed from the lexical-level address. If the referenced word is an indirect-address descriptor, its value is copied into VAR.

Instruction: EVAL (Evaluate)
Instruction Size: 8 bits
Function: Replace the descriptor on the top of the stack with the referenced word.
Initial Stack: VAR1,X
Final Stack: VAR2,X
Operands: If VAR1 is not an indirect-address descriptor or an array-element descriptor, VAR2=VAR1. Otherwise, VAR2 is the referenced word. VAR2 will never be an indirect-address descriptor or array-element descriptor; chains of descriptors are followed to locate the final item. If VAR2 has the undefined value, an error is signalled.

DATA-ACCESS AND ADDRESSING INSTRUCTIONS

Instruction: STORE

Instruction Size: 8 bits

Function: The top word on the stack is stored in the word pointed to by the second word.

Initial Stack: VAR1,VAR2,X

Final Stack: VAR1,X

Operands: VAR2 is an indirect-address descriptor, an array-element descriptor, or an array pointer. VAR1 is stored in the referenced location(s). If VAR1 is an array pointer and the referenced storage is an array pointer, the latter array is filled from the former. If VAR1 is not an array pointer, its value is copied into all elements of the latter array. If VAR1 and the target have different data types, an automatic data conversion is performed.

Notes: Data conversion rules were not explicitly specified in the original specification, but presumably they follow the rules of PL/I.

Instruction: SUBS (Subscript)

Instruction Size: 16 bits

Function: Create a descriptor to the referenced array element or subarray (cross section).

Instruction Field: N, the number of subscripts.

Initial Stack: VAR1, . . . ,VARN,VARP,X

Final Stack: VARQ,X

Operands: VAR1-VARN must be bit, fixed, float, or star words (bit and float words are converted to a fixed-point value). They represent the values of the subscripts. VARP is an array pointer or an indirect-address descriptor (or start of a chain of indirect-address descriptors) pointing to an array pointer.

If none of the subscripts is a star, VARQ is an array-element descriptor of the referenced element. If one or more of the subscripts is a star word, a descriptor of the specified subarray is built in the array descriptor memory, and VARQ is an array pointer to this descriptor. An error is signaled if any of the subscripts falls beyond the associated dimension bounds or if N is not equal to the number of dimensions of the referenced array.

Instruction: POP

Instruction Size: 8 bits

Function: The top word on the stack is discarded.

Initial Stack: VAR,X
Final Stack: X

Instruction: SWAP
Instruction Size: 8 bits
Function: The top two words on the stack are exchanged.
Initial Stack: VAR1,VAR2,X
Final Stack: VAR2,VAR1,X

Instruction: UNDEF
Instruction Size: 8 bits
Function: Push an undefined word onto the stack.
Initial Stack: X
Final Stack: VAR,X
Operands: VAR is an undefined word (u-bit in tag =1).

DATA-OPERATION INSTRUCTIONS

Instruction: ADD (Add)
Instruction: SUB (Subtract)
Instruction: MUL (Multiply)
Instruction: DIV (Divide)
Instruction: PWR (Power)
Instruction Size: 8 bits
Function: The arithmetic operation is performed on the top two stack words.
Initial Stack: VAR1,VAR2,X
Final Stack: VAR3,X
Operands: VAR1 and VAR2 must be fixed or float. If one is float, the other is automatically converted to float. The operation is performed in the form VAR3=VAR2.op.VAR1.
If VAR1 and VAR2 are array pointers, the operation is performed between corresponding array elements. The result is a newly constructed array, and VAR3 is an array pointer to the array. If one of VAR1 or VAR2 is an array pointer and the other is a scalar, a scalar operation is performed on each array element. The result is a newly constructed array, and VAR3 is an array pointer to the array.
Notes: Data conversion of a source operand does not affect the operand in storage. The conversions are only temporary during the execution of the instruction.

DATA-OPERATION INSTRUCTIONS

Instruction: CAT (Concatenate)
Instruction Size: 8 bits
Function: A string is constructed by concatenating two strings.
Initial Stack: VAR1,VAR2,X
Final Stack: VAR3,X
Operands: VAR1 and VAR2 must be bit, fixed, float, or character-string descriptors. If they are not character-string descriptors, they are converted to character strings. A new string is built in the string segment memory consisting of string VAR2 followed by string VAR1. VAR3 is a character-string descriptor. As was the case for the previous instructions, if VAR1 and/or VAR2 is an array pointer, an array operation is implied, and VAR3 is an array pointer to the new array.

Instruction: LT (Less than)
Instruction: GT (Greater than)
Instruction: LE (Less than or equal)
Instruction: GE (Greater than or equal)
Instruction: EQ (Equal)
Instruction: NE (Not equal)
Instruction Size: 8 bits
Function: The comparison operation is performed between the top two stack words.
Initial Stack: VAR1,VAR2,X
Final Stack: VAR3,X
Operands: The comparison is performed in the form VAR3 = VAR2.op.VAR1. If VAR1 and VAR2 have different types, the variable with the "lesser" type is first converted to the other type. The hierarchy of types is bit < fixed < float < character. VAR3 is a bit word (boolean value true or false). If VAR1 and/or VAR2 is an array pointer, an array-array or array-scalar comparison is done, and VAR3 is an array pointer to the new bit array.

Instruction: AND
Instruction: OR
Instruction Size: 8 bits
Function: The boolean operation is performed on the top two stack words.
Initial Stack: VAR1,VAR2,X
Final Stack: VAR3,X
Operands: VAR1 and VAR2 are converted to bit if necessary, and the

operation is performed. VAR3 is a bit word. If VAR1 and/or VAR2 is an array pointer, an array operation is implied, and VAR3 is an array pointer to the new array.

Instruction: PLUS (Unary +)
Instruction: NEG (Unary −)
Instruction Size: 8 bits
Function: The sign of the operand is left unchanged (PLUS) or inverted (NEG).
Initial Stack: VAR1,X
Final Stack: VAR2,X
Operands: VAR1 must be bit, fixed, or float. If it is bit, it is converted to fixed. If the instruction is PLUS, VAR2=VAR1. If the instruction is NEG, VAR2=−VAR1. If VAR1 is an array pointer, a new array is constructed, and VAR2 is an array pointer to the new array.

Instruction: NOT
Instruction Size: 8 bits
Function: A logical-complement operation is performed on the operand.
Initial Stack: VAR1,X
Final Stack: VAR2,X
Operands: VAR1 is converted to a bit word, if necessary, and VAR2 is its complemented value. If VAR1 is an array pointer, a new array is constructed, and VAR3 is an array pointer to the new array.

Instruction: BIT
Instruction Size: 8 bits
Function: The operand is converted to a bit word.
Initial Stack: VAR1,X
Final Stack: VAR2,X
Operands: VAR1 must be bit, fixed, or float. If VAR1 is zero, VAR2 is false; otherwise VAR2 is true.

CONTROL INSTRUCTIONS

Instruction: CALL
Instruction Size: 8 bits
Function: Suspend execution of the current program segment and execute another segment.

CONTROL INSTRUCTIONS

Initial Stack: VAR,X

Final Stack: X

Operands: VAR must be a program-segment descriptor. An entry for the new segment is created on the program segment stack.

Notes: This is not a procedure call, since no arguments are passed and no change in lexical level occurs. It is used for IF and DO statements. The CALL instruction is similar in effect to the COBOL PERFORM verb.

Instruction: DOSTORE

Instruction Size: 8 bits

Function: The top word on the stack is stored in the word pointed to by the second word.

Initial Stack: VAR1,VAR2,X

Final Stack: VAR2,X

Operands: VAR2 must be an indirect-address descriptor or an array-element descriptor.

Notes: DOSTORE is intended for the initialization of iterative DO loops, where VAR1 is the initial value of the iteration variable and VAR2 is the iteration variable. DOSTORE has the same effect as a STORE instruction, but DOSTORE is more limited in scope. If necessary, VAR1 is converted to the type of VAR2.

Instruction: DOTEST

Instruction Size: 8 bits

Function: Compare two values for a less than or equal or greater than or equal relationship.

Initial Stack: VAR1,VAR2,VAR3,X

Final Stack: VAR4,VAR1,VAR2,VAR3,X

Operands: VAR3 must be an indirect-address descriptor to an arithmetic value. VAR1 and VAR2 must be arithmetic values. VAR4, a bit value, is set to true if
   1. VAR 1 is positive and the word pointed to by VAR3 is less than or equal to VAR2.
   2. VAR1 is negative and the word pointed to by VAR3 is greater than or equal to VAR2.

Notes: DOTEST is intended for iterative DO loops, where VAR1 is the iteration increment, VAR2 is the iteration limit, and VAR3 is the iteration variable.

Instruction: CRET (Conditional Return)

Instruction Size: 8 bits

Function: Return to the previous segment if the top word on the stack has the value false.

Initial Stack: VAR,X

Final Stack: X

Operands: VAR is a bit value. If its value is one (true), execution continues with the subsequent instruction. If its value is zero, execution of the current segment is terminated and execution resumes in the previous segment.

Notes: CRET is intended for DO WHILE and iterative DO loops. In an iterative DO loop, it follows a DOTEST instruction.

Instruction: DOINCR

Instruction Size: 8 bits

Function: The word two positions below the top word on the stack is incremented by the top word.

Initial Stack: VAR1,VAR2,VAR3,X

Final Stack: VAR1,VAR2,VAR3,X

Operands: VAR3 must be an indirect-address descriptor to an arithmetic value. VAR1 must be arithmetic. VAR1 is added to the value pointed to by VAR3.

Notes: See DOTEST notes.

Instruction: CYCLE

Instruction Size: 8 bits

Function: Branch back to the first instruction in this program segment.

Initial Stack: X

Final Stack: X

Notes: CYCLE is intended for DO loops.

Instruction: GOTO

Instruction Size: 8 bits

Function: Branch to the target of a label descriptor.

Initial Stack: VAR,X

Final Stack: X

Operands: VAR must be an indirect-address descriptor to a label-constant or label-variable descriptor. If the target instruction is not in

PROCEDURE INSTRUCTIONS

the active segment, all intervening segments are terminated, and the display registers are updated if necessary.

Instruction: HALT
Instruction Size: 8 bits
Function: The program is terminated.
Initial Stack: X
Final Stack: null

Instruction: LINE
Instruction Size: 16 bits
Function: The number N is loaded into the line register.
Instruction Field: N, an 8-bit value.
Initial Stack: X
Final Stack: X

PROCEDURE INSTRUCTIONS

Instruction: ENTER
Instruction Size: 16 bits
Function: Suspend execution of the current program segment and begin execution of another segment at a possibly different lexical level.
Instruction Field: N, an 8-bit value, indicating the number of arguments.
Initial Stack: VARP,VAR1, . . . ,VARN,X
Final Stack: local storage for called procedure, VAR1, . . . ,VARN,X
Operands: VARP is an indirect-address descriptor to a program-segment descriptor. VAR1-VARN are indirect-address descriptors to the arguments being passed. ENTER verifies the number and types of the arguments against the number and types of parameters expected. It allocates storage on the data stack for the new segment, initializes the parameters as indirect-address descriptors to the arguments, places entries on the scope and program-segment stacks, and updates the display registers.
If an argument and its corresponding parameter differ in type, and if a valid data conversion is possible, the argument indirect-address descriptor is overlayed with the converted argument.
Notes: ENTER is used for procedure and function calls and for entrances into begin blocks.

Instruction: SCOPEID

Instruction Size: 16 bits

Function: It has no function when executed.

Instruction Field: M, an 8-bit address of the scope table entry corresponding to this segment.

Initial Stack: X

Final Stack: X

Notes: SCOPEID must be the first instruction of every segment that is entered with an ENTER instruction. M is used by the ENTER instruction to locate the associated scope table memory entry.

Instruction: END

Instruction Size: 16 bits

Function: Terminates execution of a segment invoked via an ENTER instruction and resumes execution of the calling segment.

Instruction Field: M, an 8-bit address of the scope table entry corresponding to this segment.

Initial Stack: VAR,X

Final Stack: VAR,X1

Operands: VAR is the returned value. X1 represents the stack contents at the beginning of the corresponding ENTER instruction, minus VARP (the indirect-address descriptor to the program-segment descriptor). END restores the scope and program segment stacks to their previous state and updates the display registers.

Instruction: PARAM (Parameter)

Instruction Size: 16 bits

Function: If the operand is a procedure, execution of the current program segment is suspended, and execution of the procedure is started. If the operand is not a procedure, the instruction does nothing, and the stack is unchanged.

Instruction Field: P, an 8-bit number.

Initial Stack: VAR1,VAR2,X

Final Stack: local storage for called procedure, VAR2,X

Operands: VAR1 is an indirect-address descriptor to an argument. P indicates that VAR1 represents the Pth parameter to VAR2 (VAR2 is an indirect-address descriptor to a program-segment descriptor). If the argument is a parameterless function that returns a value (i.e., if

VAR1 points to a program-segment descriptor), VAR1 is entered as if an ENTER instruction had been executed.

Notes: PARAM is used to determine if a procedure argument is being passed to a nonprocedure parameter, meaning that the argument must be evaluated. PARAM instructions are generated by the compiler preceding each ENTER instruction.

Instruction: LFUNC (Linkage to Pseudofunction)
Instruction Size: 16 bits
Function: Suspend execution of the current program segment and begin execution of another segment representing a pseudovariable.
Instruction Field: N, an 8-bit value indicating the number of arguments.
Initial Stack: VARR,VARF,VAR1, . . . ,VARN,X
Final Stack: unspecified
Operands: VARR is an indirect-address descriptor to a value, which is converted to a character string. VARF is an indirect-address descriptor to a fixed-point value. The built-in function corresponding to this value is invoked. VAR1-VARN are indirect-address descriptors to the arguments being passed.
Notes: LFUNC is used to represent pseudovariables (OUTPUT and SUBSTR) on the left side of assignment statements. VARR represents the value of the right side of the assignment statement. LFUNC is similar in effect to a "supervisor call."

ARRAY-STORAGE INSTRUCTIONS

Instruction: ALLOC (Allocate)
Instruction Size: 16 bits
Function: Allocate storage for an array.
Instruction Field: D, an 8-bit value indicating the number of dimensions.
Initial Stack: VARA,VARL1,VARU1, . . . ,VARLD,VARUD,X
Final Stack: X
Operands: VARA is an indirect-address descriptor to an array pointer. VARL1-VARUD are fixed-point words representing the lower and upper subscript bounds for each dimension. ALLOC allocates space for the array elements (in the upper portion of the data stack), creates a descriptor in the array descriptor memory, and stores the address of the descriptor in the array pointer. If there was a prior version of the

array (i.e., if the array pointer was not undefined), the new array descriptor points to the previous descriptor.

Instruction: FREE
Instruction Size: 8 bits
Function: Frees the storage of an array or a version of an array.
Initial Stack: VARA,X
Final Stack: VARA,X
Operands: VARA is an indirect-address descriptor to an array pointer. The associated descriptor in the array descriptor memory and the space occupied by the elements are freed. If there is a prior version of the array (the array descriptor points to another array descriptor), the array pointer is updated to point to the next descriptor. If not, the array pointer is marked as undefined.

III
A High-Level-Language Architecture

8 | System Architecture of the Symbol System

The second case study of an advanced computer architecture is an operational system. The system is the SYMBOL system [1–21], developed by the Fairchild Camera and Instrument Corporation. Only one system was constructed, and it is in use at Iowa State University.

The SYMBOL system was the result of a major research project to reexamine the traditional division between hardware and software. The goals of the research were

1. Design a computing system with substantial performance increases over conventional systems, along with a reduction in the cost of computing, by directly implementing a high-level language and virtual-storage, time-sharing, operating system in hardware.
2. Rather than orienting the system to traditional "number-crunching" applications, design it to assist the programmer in working with non numeric data and elaborate data structures.
3. Develop new hardware design and construction techniques.
4. Demonstrate the above with a full-scale working system.

Based on the categories defined in Chapter 3, SYMBOL can be classified as a type-B high-level-language architecture. Its instruction set is a one-for-one reverse-Polish representation of the system's programming language, the SYMBOL Programming Language (SPL), and the compilation process is performed by the machine rather than by software.

An initially startling aspect of SYMBOL is the first goal; the phrase "directly implementing ... in hardware" means that SYMBOL contains virtually no software (it does contain some software, but the software performs only a few auxiliary functions, and the system can be operated with no software), and it contains no microcode; all the system's functions are implemented as sequential logic networks. For instance, the "compiler" is a hardware sequential logic network, the memory-allocation and page-replacement algorithms are sequential logic networks, the functions that interpret end-user terminal commands are sequential logic networks, and so on. Although this may seem initially to be the most unusual aspect of SYMBOL, it is not emphasized because it also turns out to be one of the shortcomings of the system, and the same objectives could have been achieved through the use of microprogramming.

To the user, SYMBOL appears to be a terminal-oriented, time-sharing, virtual-storage system supporting the SPL language. The internal organization and machine language of the system are totally transparent to the user. The user debugs his programs in terms of the programming language and never sees a traditional machine address or storage dump. The only other property of SYMBOL visible to the user is that it appears to be an extremely fast system.

SYSTEM CONFIGURATION

Internally, SYMBOL is a multiprocessing system, but it is a unique application of multiprocessing in two respects. First, the user is totally unaware that it is a multiprocessing system. Second, each processor is dedicated to a specific computing function. For instance, one processor is dedicated to compiling SPL programs, another contains the dispatching and paging algorithms, another is dedicated to processing terminal commands, another to storage management, another to user-program execution, and so on.

The processors and their interconnections are illustrated in Figure 8.1. The number in the lower corner of each box is a measure of the relative complexity of the processor; the number indicates the number of circuit boards in the processor. A circuit board contains approximately 200 integrated circuits, or approximately 800 gates.

The functions of each of the processors are discussed in Chapters 9 and 10, but a brief description is appropriate at this point. The processors each have a queue of requests needing service, and all processors can be active simultaneously. A user's program is passed from processor to processor as different activities must occur, but a particular user

SYSTEM CONFIGURATION

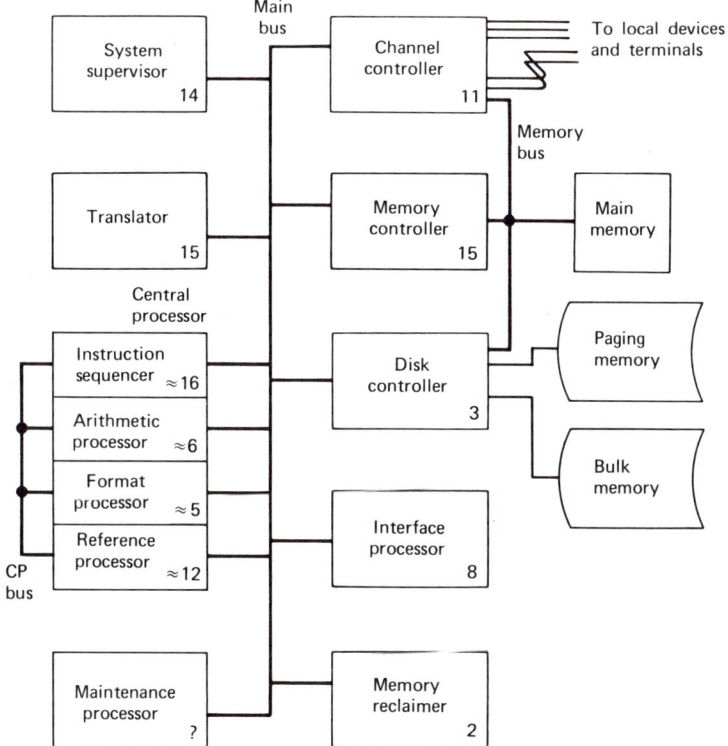

Figure 8.1   SYMBOL processor configuration.

program is never active in more than one processor at any time. Thus, to the user, the system appears to be a single-processor system.

The *channel controller* acts as an intermediary between the 32 I/O channels (each of which represents a terminal or peripheral device) and their associated buffers in main memory. The *interface processor* also performs functions related to input/output. When an input buffer is full, the interface processor copies it into the appropriate location in a user's virtual memory. The interface processor also acts as the I/O subsystem in a conventional operating system; when a user program executes an input or output instruction, the request is handled by the interface processor. The interface processor also functions as the "command-handling" subsystem of a conventional time-sharing system; for instance, if a terminal user is in the process of editing a source program, the editing commands are handled by the interface processor, leaving the other processors free to perform work on behalf of other users.

The *translator* is an independent processor whose sole function is compilation. That is, the only function the translator is capable of performing is taking SPL programs from its input queue and compiling them into object programs.

The only function of the *central processor* is the execution of object programs. As indicated in Figure 8.1, the central processor actually consists of four separate processors. However, these four processors operate serially, implying that only one program is being executed at any instant in time.

Four of the processors are service processors; rather than independently performing system functions, they perform functions at the request of other processors. The *memory controller* services other processors that wish to fetch or store information in memory. The memory controller gives the other proccesors a high-level abstract view of memory: from the point of view of the other processors, memory has the appearance of being an infinite number of infinite-length lists. This concept of memory and the associated processes of memory management are the major unique characteristics of SYMBOL; they are discussed in more detail in Chapter 10.

The second service processor is the *memory reclaimer*. Because of the methods used by the memory controller to map the logical view of memory (list structures) into a virtual memory, and virtual memory into main memory, a request by a processor to free a list is potentially a time-consuming operation. Rather than processing such a request serially and tying up the requesting processor and the memory controller for an inordinate amount of time, the memory controller simply marks the associated storage as unavailable and puts the storage in a queue for later reclamation by the memory reclaimer processor.

The third service processor is the *system supervisor*, which performs functions analogous to process dispatching and memory paging in conventional operating systems. The system supervisor manages the queus for the other processors and schedules paging operations when one of the processors encounters a page fault when dealing with the memory controller.

The last service processor is the *disk controller*. The disk controller acts as a queue-driven I/O channel; it processes requests in its queue and transfers information between main memory and the secondary storage devices.

The remaining processor is the *maintenance processor*. The maintenance processor does not have an active role in normal system operations. It is used for system debugging and to monitor or trace specific events on the main bus interconnecting the processors.

JOB FLOW THROUGH THE SYSTEM

All the processors communicate via the main bus as shown in Figure 8.1. The main bus has three purposes:

1. To pass data between any processor and the memory controller.
2. To pass control information between any processor and the system supervisor.
3. To pass data among the four processors in the central processor.

When sending requests to the memory controller, the processors use the bus on a priority basis; since memory is the critical resource, these priorities are an approximate measure of the relative priorities of the processors. The memory priorities from highest to lowest are the system supervisor, interface processor, central processor, translator, and memory reclaimer. The channel controller and disk controller are not included in this list because they do not use the memory controller; they share a direct bus to main memory, as shown in Figure 8.1. A third bus is used in the central processor to pass control information among its four processors.

JOB FLOW THROUGH THE SYSTEM

The easiest way to understand the dynamics of the system is to study the flow of a user's job from processor to processor during different stages of its processing. To do this, several simplfying assumptions are made. We assume that no page faults occur (i.e., all the necessary data are in main memory). Also, each processor time-slices the work on its queue, but we assume that no end-of-time slice conditions occur. Paging and time-slicing are discussed separately in Chapter 10.

The primary SYMBOL terminal has, in addition to a normal keyboard, a set of keys representing system commands. Examples of the keys include LOAD, RUN, PAUSE, CONTINUE, SEARCH, REMOVE, FORWARD, BACKWARD, and DISPLAY (the last five are editing functions). Initially we assume that the user is in load mode, a text entry and editing mode. Only four processors are used in this mode: the system supervisor, channel controller, interface processor, and memory controller. These four processors are likely to be performing work on behalf of other users, and the translator and central processor are working independently on behalf of other users.

If the terminal user is entering a text line (e.g., a statement of the program that he is writing), the channel controller receives each character and places it in a line buffer associated with the terminal in a fixed area of main memory. When the buffer is full, the channel controller sends a buffer-full signal to the system supervisor and switches to a back-up buffer. The system supervisor puts a notice in the queue for the interface processor, telling it to transfer the contents of this buffer into the user's transient working area. When the interface processor encounters this queue entry, it transfers the buffer to the working area by transmitting the appropriate requests to the memory controller.

If the user depresses one of the command keys (e.g., he depresses the BACKWARD key to move the current-line pointer in the working area back one line), a control character is received by the channel controller. Seeing that it is a control character, the channel controller does not place it into the buffer, but it sends it to the system supervisor. If the control character represents an editing operation (e.g., BACKWARD), the system supervisor places the request in the interface processor's queue. When the interface processor encounters this queue entry, it performs the editing operation using the memory controller.

A second scenario occurs when the user depresses the RUN key, which means "compile the program in my transient working area and execute it." As before, the channel controller sees the control character and sends it to the system supervisor. The system supervisor builds an entry in the translator's queue. Eventually the translator encounters this queue entry and begins compiling the program.

The translator is analogous to both a compiler and a linkage editor. If it encounters an unresolved reference to a procedure, it fetches the procedure from one of the program libraries in bulk storage. Eventually, by communicating with the memory controller, the translator builds two lists in memory: an object-code string and one or more name tables.

The translator is so fast that a decision was made to store all programs (i.e., the programs in the libraries) in their source form. If I/O operations and page faults are overlooked, the average effective speed of the translator is the compilation of 75,000 statements per minute. (In an optimized version of SYMBOL that was designed but never built, the translator was optimized to an estimated 300,000 statements per minute [2].)

When the translator has completed the compilation, it notifies the system supervisor. The system supervisor places an entry on the central processor's queue, and the central processor eventually begins execution of the program. The central processor uses the memory controller to access the object string and the name tables and to allocate storage for, and alter the values of, the program's variables. If the central processor encounters an I/O instruction in the object string, it gives the request to the system supervisor, which adds the request to the interface processor's queue.

A possible reaction to the SYMBOL system is that all this interprocessor communication must lead to considerable overhead. A good way to measure this, and a good way to compare SYMBOL to a conventional time-sharing system, is to analyze the time needed to compile and execute a one-statement "do nothing" program. Assuming only a single user on the system, the elapsed time from receipt of the RUN command to the completion of execution of the program is 316 microseconds. Comparing this to a conventional system, it is likely that the conventional system needs more than 315 microseconds to analyze the compile command, let alone begin executing its compiler. This 316-microsecond time is even more phenomenal when one realizes that the limiting factor is the relatively slow ferrite-core memory in SYMBOL with a cycle time of 4 microseconds per 64-bit word.

THE SYMBOL PROGRAMMING LANGUAGE

The SYMBOL Programming Language (SPL) is a simple, general-purpose, high-level programming language. Since the machine language of the central processor is essentially a one-for-one reverse-Polish representation of SPL, it is worthwhile to study SPL before examining the architecture of the central processor.

In comparison to other languages, SPL can be viewed as a mixture of ideas in APL, PL/I, and LISP. It is similar to APL in that it is a typeless language (there are no declaration statements for variables, and the attributes of variables can change during execution), and there are only a few basic data types and structures. SPL is similar to PL/I in terms of syntax, block structures, and ON-units, and it is similar to LISP in that its primary data structure is a list.

SPL contains only two types of data: scalars and lists (called structures). A scalar is a string of characters of arbitrary length. There are 20 operators defined for scalar operands. As an illustration, the program segment

```
A = 22.3;
B = |abcD$| ;
C = A join B;
OUTPUT C;
```

yields the result 22.3abcD$ printed on the terminal. The field mark "|" is used to delimit literal scalar values, but its use is optional for scalars that have the syntax of numeric values. For ease of typography, the equal sign is used here for the assignment operator, but in the language an arrow is used.

SPL contains numeric operators that are defined on scalar operands that have the syntax of numbers. Examples of such scalars are

0.5   3.14159EM
486   32.2EX

Although in general scalars can have an unlimited length (limited only by the amount of memory available), numeric scalars have a maximum length of 99 digits. If the value of a number is less than 10^{-10} or greater than 10^{10}, it is stored internally in mantissa/exponent form. A mantissa can contain from 1 to 99 digits, and an exponent can contain 1 or 2 digits. All arithmetic is done in base 10.

Storing numerical values in a variable number of digits can lead to excessive processing time and storage problems, because the result of

A = 1/3;

is the 99-digit number .3333.... To overcome this, a pseudovariable LIMIT and the data attributes *exact* (EX) and *empirical* (EM) were added to the language. LIMIT can be used by the programmer to set the size of results of subsequent numerical operations. For instance, the program segment

LIMIT = 5;
A = 1/3;

assigns A the value .33333EM. EM is an attribute stored with the value of A; it indicates that A is not an exact value. The rule for numerical operations is that the length of the result will be the lesser of (1) the current value of LIMIT and (2) if one or more of the operands is empirical (i.e., has the EM attribute), the length of the shortest empirical operand. If nonzero digits of the result are truncated because of this, or if one of the operands was empirical, the result is marked as empirical. This is illustrated by the program segment

```
LIMIT = 5;
A = 1/3;              result is .33333EM
B = 1.0;              result is 1.0000EX
LIMIT = 10;
C = B/A;              result is 3.0000EM
D = B;                result is 1.000000000EX
E = D * A;            result is .33333EM
```

The program can determine if an operation resulted in a limited-precision result by testing the boolean pseudovariable LIMITED.

THE SYMBOL PROGRAMMING LANGUAGE 105

SPL also contains a set of boolean operators that are defined for scalar operands containing only 0s and 1s.

Structures

A structure is an ordered list whose elements can be scalars or structures. A structure may be used as an array, but it is a more powerful concept than an array because its size and shape can vary dynamically and its elements need not have identical attributes. A structure literal is delimited by group marks (< and >), and its elements are separated by field marks. This program segment illustrates some of the properties of structures:

```
A = <100|200|300>;
OUTPUT A[1];               result is 100
A[4] = |WXYZ|;
OUTPUT A;                  result is < 100|200|300|WXYZ >
A[1] = <101|102>;
OUTPUT A[1, 2];            result is 102
OUTPUT A;                  result is < <101|102>|200|300|WXYZ >
B = A;
OUTPUT B;                  result is <<101|102>200|300|WXYZ>
B[1] = 100;
OUTPUT B;                  result is < 100|200|300|WXYZ >
```

Any reference to a nonexistent structure element brings that element into existence. For instance, the statement
    A[6] = B;
moves structure B into the sixth element of A, and A[5] has a null value.

Program Structure and Control

SPL contains IFTHENELSE and GOTO statements. Although an iteration statement (LOOP) was planned, it was not implemented in SYMBOL.

SPL contains procedures and functions. The argument-transmission method for both is transmission by name, as is indicated by the program segment

```
A = 1;
B = 2;
CALL W(A,A+B);
```

```
PROCEDURE W(X,Y);
OUTPUT Y;           result is 3
X = X+1;
OUTPUT Y;           result is 4
END
```

Like ALGOL and PL/I, SPL is a block-structured language with scope-of-variable-name rules among nested procedures. Unlike ALGOL and PL/I, however, the default condition in SPL is that an identifier is local to the procedure in which it is used. An identifier can be specified as synonymous with the identical identifier in the enclosing procedure by use of a GLOBAL statement. The statement

```
GLOBAL X;
```

makes X in this procedure synonymous with identifier X in the enclosing procedure.

SPL also contains ON units, which behave as procedures except for the way in which they are invoked. If the ON statement lists one or more variable names, the ON unit is entered immediately after an assignment to one of the variables. If the ON statement lists one or more statement labels, the ON unit is entered immediately before a transfer (GO TO) to one of the labels. If the ON statement lists one or more procedure names, the ON unit is entered immediately before a call to one of the procedures. The ON statement can also specify INTERRUPT, which causes the ON unit to be entered when a terminal interrupt occurs.

Table 8.1 lists the statement types and the operators in SPL.

Table 8.1 SPL Language Summary

Statement Types		Operators	
INPUT	SWITCH	+	BEFORE
OUTPUT	ON	−	SAME
GO TO	DISABLE	*	AFTER
IF	ENABLE	/	JOIN
Assignment	PROCEDURE	ABS	AND
CALL	BLOCK	GREATER	OR
PAUSE	RETURN	GTE	NOT
CONTINUE	GLOBAL	EQUALS	FORMAT
LINK	SYSTEM	NEQ	MASK
TRAP	NOTE	LTE	IN
Initial value		LESS	

As an illustration of an SPL program, these are the second Student-PL program from Chapter 6 and the equivalent SPL program.

```
XYZ: PROCEDURE;            B = 4;
DECLARE A(*) FLOAT;        I=1;
DECLARE (B,C) FIXED;       L: IF I LTE B THEN
B=4;                          A[I]=1;
ALLOCATE A (1:B);             I=I+1;
A=1;                          GO TO L;
C=B+B*B;                      END
CALL ZZZ (B);              C=B+B*B;
A(B)=C;                    CALL ZZZ (B);
   ZZZ: PROCEDURE (M);     A[B]=C;
   DECLARE M FIXED;           PROCEDURE ZZZ (M);
   C=M+M;                     GLOBAL C;
   END;                       C=M+M;
END;                          END
```

REFERENCES

1. G. D. Chesley and W. R. Smith, "The Hardware-Implemented High-Level Machine Language for SYMBOL," *Proceedings of the 1971 Spring Joint Computer Conference.* Montvale, N. J.: AFIPS, 1971, pp. 563–573.
2. R. Rice and W. R. Smith, "SYMBOL—A Major Departure from Classic Software Dominated von Neumann Computing Systems," *Proceedings of the 1971 Spring Joint Computer Conference.* Montvale, N.J.: AFIPS, 1971, pp. 575–587.
3. B. E. Cowart, R. Rice, and S. F. Lundstrom, "The Physical Attributes and Testing Aspects fo the SYMBOL System," *Proceedings of the 1971 Spring Joint Computer Conference.* Montvale, N. J.: AFIPS, 1971, pp. 589–600.
4. W. R. Smith et al., "SYMBOL—A Large Experimental System Exploring Major Hardware Replacement of Software," *Proceedings of the 1971 Spring Joint Computer Conference.* Montvale, N. J.: AFIPS, 1971, pp. 601–616.
5. M. A. Calhoun, "SYMBOL Hardware Debugging Facilities," *Proceedings of the 1972 Spring Joint Computer Conference.* Montvale, N. J.: AFIPS, 1972, pp. 359–368.
6. R. Rice, "The Hardware Implementation of SYMBOL," *Digest of the Sixth Annual IEEE Computer Society International Conference.* New York: IEEE, 1972, pp. 27–29.
7. W. R. Smith, "System Supervision Algorithms for the SYMBOL Computer," *Digest of the Sixth Annual IEEE Computer Society International Conference.* New York: IEEE, 1972, pp. 21–25.
8. R. J. Zingg and H. Richards, Jr., "Operational Experience with SYMBOL," *Digest of the Sixth Annual IEEE Computer Society International Conference.* New York: IEEE, 1972, pp. 31–32.
9. T. A. Laliotis, "Implementation Aspects of the SYMBOL Hardware Compiler," Pro-

ceedings of the First Annual Symposium on Computer Architecture. New York: IEEE, 1973, pp. 111–115.
10. H. Richards, Jr. and R. J. Zingg, "The Logical Structure of the Memory Resource in the SYMBOL-2R Computer," Proceedings of the ACM-IEEE Symposium on High-Level-Language Computer Architecture. New York: ACM, 1973, pp. 1 10.
11. J. W. Anderberg and C. L. Smith, "High-Level Language Translation in SYMBOL 2R," Proceedings of the ACM-IEEE Symposium on High-Level-Language Computer Architecture. New York: ACM, 1973, pp. 11–19.
12. P. C. Hutchison and K. Ethington, "Program Execution in the SYMBOL 2R Computer," Proceedings of the ACM-IEEE Symposium on High-Level-Language Computer Architecture. New York: ACM, 1973, pp. 20–26.
13. H. Richards, Jr. and C. Wright, Jr., "Introduction to the SYMBOL 2R Programming Language," Proceedings of the ACM-IEEE Symposium on High-Level-Language Computer Architecture. New York: ACM, 1973, pp. 27–33.
14. H. Richards, Jr., "SYMBOL II-R Programming Reference Manual," ISU-CCL-7301, Cyclone Computer Laboratory, Iowa State University, Ames, Iowa, 1973.
15. A. C. Bradley, "An Algorithmic Description of the SYMBOL Arithmetic Processor," ISU-CCL-7305, Cyclone Computer Laboratory, Iowa State University, Ames, Iowa, 1973.
16. R. J. Zingg and H. Richards, Jr., "SYMBOL: A System Tailored to the Structure of Data," ISU-CCL-7302, Cyclone Computer Laboratory, Iowa State University, Ames, Iowa, 1973.
17. W. E. Jones, "The Role of the Interface Processor in the SYMBOL IIR Computer System," NSF-OCA-GJ33097-CL7304, Cyclone Computer Laboratory, Iowa State University, Ames, Iowa, 1973.
18. M. C. Dakins, "Nonnumeric Processing in the SYMBOL-2R Computer System," NSF-OCA-GJ33097-CL7402, Cyclone Computer Laboratory, Iowa State University, Ames, Iowa, 1974.
19. J. W. Anderberg, "Source Program Analysis and Object String Generation Algorithms and Their Implementation in the SYMBOL 2R Translator," NSF-OCA-GJ33097-CL7410, Cyclone Computer Laboratory, Iowa State University, Ames, Iowa, 1974.
20. T. A. Laliotis, "Architecture of the SYMBOL Computer System," in Y. Chu, Ed., High-Level Language Computer Architecture. New York: Academic, 1975, pp. 110–185.
21. H. Richards, Jr. and A. E. Oldehoeft, "Hardware-Software Interactions in SYMBOL-2R's Operating System," Proceedings of the Second Annual Symposium on Computer Architecture. New York: IEEE, 1975, pp. 113–118.

EXERCISES

8.1 Why is the "all hardware" implementation of SYMBOL a shortcoming?

8.2 What is the meaning of the SPL expression X[2,3,4]?

EXERCISES

8.3 Is the result of the assignment statement A=X[2,3,4] a scalar or a structure?

8.4 The section on SPL structures contained a set of statements ending with A[6]=B. Draw a diagram of structure A after the execution of this statement.

9 | Computer Architecture of the SYMBOL System

In relation to the architectural categories in Chapter 3, SYMBOL can be classified as a type-B high-level-language architecture. The instruction set of the central processor is a one-for-one reverse-Polish representation of SPL; the function of the translator processor is to translate SPL programs into this representation. In relation to the concepts of Chapter 4, the central processor's architecture employs self-identifying data, descriptors of data objects, subroutine management, dynamic storage management, variable-size data, and decimal arithmetic.

The instruction-set architecture of the central processor is the least-documented aspect of the system. No definitive document of the instruction set exists; the material in this chapter was pieced together by analyzing several object programs. Because of this, the material in this chapter is subject to inaccuracies, although it is accurate enough to present a flavor of the central processor's architecture. Also, the entire instruction set is not discussed; the instruction set is discussed by example, and only those aspects of the instruction set applicable to the example are discussed.

REPRESENTATION OF DATA

The central processor has two representations of scalars and structures: an external form and an internal form. The external form is the form in which scalars and structures are represented in memory when they are communicated between the central processor and other processors (e.g., the translator and the interface processor). Data may also be stored

REPRESENTATION OF DATA 111

in memory in an internal form for more efficient processing. Since the external form is the only representation visible from outside the central processor (i.e., the internal forms are not part of the central processor's computer architecture), the internal forms are not discussed in this chapter; they are discussed in Chapter 10.

The external form of a scalar is a *string*. The internal forms are *string* and *numeric*. A string is a variable-length linear list of 8-bit characters stored in an integral number of 64-bit words. A string is delimited by string-start (F5) and string-end (F6) characters. The value of the string is represented in the ASCII code. The high-order bit in the last character in the last word of a string must be on; the convention is to use the F6 character. As an example, the scalar "DOG" is represented in a string as

F5 44 4F 47 F6 XX XX F6

where "X" represents a "don't-care" character. The scalar "−289.143EM" is represented in the string occupying two words:

F5 2D 32 38 39 2E 31 34
33 45 4D F6 XX XX XX F6

The external form of a structure is called the *linear representation*. The internal form is called the *normal representation*. In the linear representation, an entire structure is delimited by two words beginning with a left-supergroup (FD) and right-supergroup (FF) character. Each substructure (a structure element that is a structure rather than a scalar) is delimited by two words beginning with a left-group (FC) and right-group (FE) character (corresponding to group marks < and > in SPL). Each scalar element is stored as a string. As an example, the structure

<101 | <201|202>|ABCD>

has the linear representation

FD XX XX XX XX XX XX XX
F5 31 30 31 F6 XX XX F6
FC XX XX XX XX XX XX XX
F5 32 30 31 F6 XX XX F6
F5 32 30 32 F6 XX XX F6
FE XX XX XX XX XX XX XX
F5 41 42 43 44 F6 XX F6
FF XX XX XX XX XX XX XX

At this point the reason for separate internal representations should be apparent. For instance, although the representation of the structure is ideal from the point of view of the interface processor (i.e., to allow it to format the structure for printing on the terminal), it is not an ideal form for use with subscripting operations, because a subscript operation would require a sequential search through each word. Hence the purpose of separate internal forms is efficient processing by the central processor.

The string, group, and supergroup delimiters, along with the data, are roughly equivalent to the concept of tags discussed in Chapter 4. The reason for including the data as part of the self-identifier is that the value of the data defines its type (i.e., the allowable operations). For instance, if a string consists of only 0s and 1s, boolean, arithmetic, and string operations are allowed. If a string obeys the syntax of numbers, arithmetic and string operations are allowed; if not, boolean and arithmetic operations are prohibited on the string by the machine and will result in a program error.

THE NAME TABLE

An SPL program is presented to the central processor as an object-code string and one or more name tables. (This use of the word "string" has nothing to do with the concept in the previous section.) The translator produces one name table per block in the source program; the name table serves roughly the same purpose as the symbol table and scope table in the Student-PL machine of Part II.

The name table consists of a *block control word* followed by entries for each identifier (e.g., statement label, variable, procedure) in the block. The format of a name table is shown in Figure 9.1, and the interpretation of the first word (the block control word) is defined in Table 9.1. The meanings of most of the fields should be clear. Bit 4 indicates that this block has the authority to execute instructions that are restricted to system programs.

The entries in the name table consist of two or more words. The first part of an entry is the name of the identifier, stored as a scalar string in one or more words. The name is used only when an error must be reported to the user. The second part is an *identifier control word*, which is analogous to the concept of descriptors discussed in Chapter 4. Oversimplifying a bit, operand addresses in the object program point to identifier control words, and identifier control words point to the values of the operands. The interpretation of the identifier control word is shown in Tables 9.2 and 9.3.

THE NAME TABLE

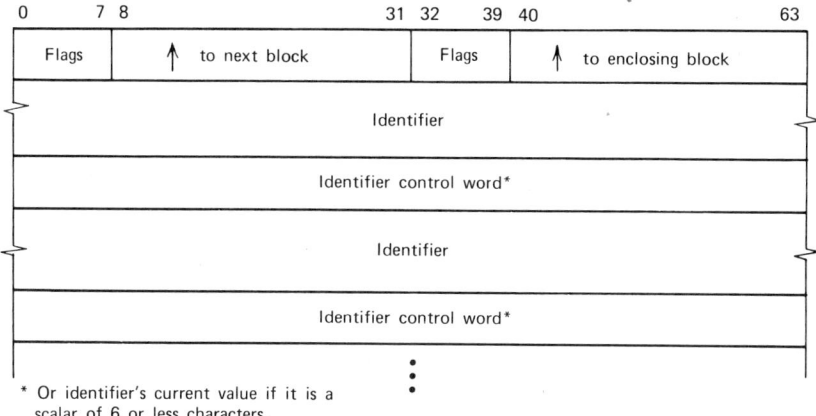

* Or identifier's current value if it is a scalar of 6 or less characters.

Figure 9.1 SYMBOL name table.

Note that the identifier control word has two basic formats. If the first 4 bits are on, the identifier's value is stored directly in the identifier control word. Otherwise, the identifier control word has the format shown in Table 9.2, and, if the identifier currently has a value, bits 8–31 contain the virtual address of the first word of the value. If a machine instruction that assigns a value to an identifier is being processed, the central processor first examines the identifier control word and the

Table 9.1 Block Control Word

Bits	Meaning
0	Always on (indicates a control word)
1	Always on (indicates a block control word)
2	Always off
3	Unused
4	Privileged procedure
5	Always on (used only during compilation)
6	A forward link (bits 8–31) is present
7	A backward link (bits 40–63) is present
8–31	Address of name table of next block in program
32–36	Unused
37	Block is active
38	Block has recursed
39	Unused
40–63	Address of name table of enclosing block

Table 9.2 Identifier Control Word

Bits	Meaning
0	Always on (indicates a control word)
1	Usually off (indicates an identifier control word) If on, see note below[a]
2	Last identifer control word in name table[a]
3–4	00—normal local variable
	01—identifier is global
	10—initialized local variable or label, procedure, or parameter
	11—not understood
5	Not understood
6	Structure or structure element
7	Extended flags (32–37) are present
8–31	See Table 9.3
32	ON enabled[b]
33	Unused[b]
34	Identifier is a label[b]
35	Identifier is a procedure[b]
36	Identifier is a parameter[b]
37	An ON reference to the identifer exists[b]
38–39	Unused[b]
40–63	See Table 9.3

[a] If bits 0–3= 1111, they and the remainder of the bits have a new meaning; they indicate that the value of the identifer is stored in bits 0–63.

[b] The meaning if bit 7 is on. If bit 7 is off, bits 32—39 are used to hold the last-used subscript (see Chapter 10).

value. If bit 7 is off (indicating that the identifier is a variable with no associated ON unit) and the value being assigned is a scalar of six characters or less, the value is stored in the identifier control word. If not, space for the value is allocated through the use of the memory controller, and the assigned address is placed in bits 8–31 (unless the variable is global or a parameter, in which case the address is placed in the identifier control word of the associated variable).

As an example, if a local variable is assigned the value 123, the resultant identifier control word is

F5 31 32 33 F6 XX XX F6

If the local variable is assigned the value −123.56789, the resultant identifier control word might be

80 00 77 77 XX XX XX XX

THE NAME TABLE

Table 9.3 Address Fields in Identifier Control Word

If Identifier Is	Then
A local variable (bits 3–4 = 00 or 10)	bits 8–31: start address of value (if any) bits 40–63: address of associated ON block (if any)
A local structure variable with no associated ON block (bits 3–4 = 00 or 10, bit 6 = 1, bit 7 = 0)	bits 8–31: start address of value (if any) bits 32–39: value of last-used subscript bits 40–63: start address of last-used component
A global variable (bits 3–4 = 01)	bits 8–31: address of identifer control word of corresponding variable in enclosing block bits 40–63: address of associated ON block (if any)
A parameter (bits 7,36=1)	bits 8–31: address of identifer control word or object code of corresponding argument (if any) bits 40–63: address of associated ON block (if any)
A label (bits 7,34=1)	bits 8–31: address of object code bits 40–63: address of associated ON block (if any)
A procedure (bits 7,35 =1)	bits 8–31: address of object code bits 40-63: address of associated ON block (if any)

assuming that location 7777 was assigned for the storage of the value.

This program is used to illustrate the production of the name tables and object-code string by the translator.

```
B = 4;
I = 1;
L: IF I LTE B THEN
   A[I] = 1;
   I = I + 1;
   GO TO L;
END
```

```
C = B + B * B;
CALL ZZZ(B);
A[B] = C;
  PROCEDURE ZZZ (M);
  GLOBAL C;
  C = M + M;
  END
```

Figure 9.2 shows the resultant name tables. The number to the left of each entry shows the entry's virtual address. Although a name table

Virtual address					Identifier
2220	C6	00 22 88	00	00 00 00	
2221	F5	42 F6 XX	XX	XX XX F6	B
2222	80	00 00 00	XX	XX XX XX	
2223	F5	49 F6 XX	XX	XX XX F6	I
2224	80	00 00 00	XX	XX XX XX	
2225	F5	4C F6 XX	XX	XX XX F6	L
2226	91	00 30 08	20	XX XX XX	
2227	F5	41 F6 XX	XX	XX XX F6	A
2270	80	00 00 00	XX	XX XX XX	
2271	F5	43 F6 XX	XX	XX XX F6	C
2272	80	00 00 00	XX	XX XX XX	
2273	F5	5A 5A 5A	F6	XX XX F6	ZZZ
2274	B1	00 30 72	10	XX XX XX	

2288	C5	00 00 00	00	00 22 20	
2289	F5	4D F6 XX	XX	XX XX F6	M
228A	91	00 00 00	08	XX XX XX	
228B	F5	43 F6 XX	XX	XX XX F6	C
228C	A8	00 22 72	XX	XX XX XX	

Figure 9.2 Name tables associated with the example.

is perceived by the translator and central processor as being stored in consecutive locations in a list, the entries are not necessarily consecutive in virtual storage. However, this is transparent to all but the memory controller and is associated with the method in which the memory controller maps its list-structure concept of memory into a sequential virtual memory. This concept is discussed in Chapter 10.

Since the attributes and sizes of variables can vary dynamically, the translator assigns no storage for the variables. Hence bits 8–31 of all the local variables and parameters are initially zero. Identifiers L and ZZZ are a label and procedure, respectively; thus their identifier control words point to locations within the object-code string. The object-code string for this program is developed in the next section. (The code for label L will start at location 3008, and the code for procedure ZZZ will start at location 3072.) Also note that C is global in the second name table; thus the identifier control word for C points to the identifier control word for C in the name table for the outer procedure.

THE OBJECT-CODE STRING

The instruction set of the central processor is stack oriented and conceptually similar to that of the Student-PL machine. The key differences are in the form of operand addressing (rather than using lexical-level addressing, SYMBOL instructions address their operands via the identifier control words), storage compactness (the SYMBOL instruction set is less efficient in terms of space), and the lack of array operations (except for the assignment operation) in SYMBOL.

Each SYMBOL instruction begins with an 8-bit operation code. Depending on the operation code, the remainder of the instruction has one of four formats:

1. No operand. The instruction length is 32 bits; the last 24 bits are ignored.
2. A data operand. The instruction length is 32 bits; the last 24 bits represent a virtual address of an identifier control word.
3. An instruction operand. The instruction length is 32 bits; the last 24 bits represent a virtual address of an instruction.
4. A literal operand. The instruction length is variable, but it must be an integral number of words, and the opcode must be on a word boundary. The operation code is the string-start, left-supergroup, or left-group character; the operand is the remainder of the string or linear structure.

In general, instructions are packed two to the word, and operation codes must begin on word or halfword boundaries. Instructions that do not begin on word boundaries are not addressable; this leads to the need to insert no-operation (no-op) instructions occasionally where an instruction must be aligned on a word boundary so that it may be addressed by another instruction. Also, for reasons unknown to the author, certain instruction types *must* begin on word (64-bit) boundaries, and certain other instruction types *must not* begin on word boundaries, again necessitating the generation of no-op instructions.

The object code for the previous SPL program is used as an example, along with the name table in Figure 9.2. In a manner analogous to the use of the LINE instruction in the Student-PL machine, the object code in SYMBOL points to its related statements in the source language for debugging purposes. In SYMBOL this is done by storing the source program in memory during the execution of the object program. Figure 9.3 represents the source program as stored in memory.

The first third of the object program is illustrated in Figure 9.4. The first column is the virtual address of each word. The op-code and operand addresses are in the second and third columns. As before, "X" is a "don't-care" character. The fourth column contains comments about each instruction. Consecutive machine instructions are not necessarily stored in consecutive virtual-storage locations. This is transparent to the central processor and only apparent to the memory controller; to the central processor, the instructions are stored consecutively in a linear list.

Virtual address	Word contents
1000	B=4; I=
1001	1; L: IF
1002	I LTE B
1003	THEN A[I
1004	]=1; I=I
1005	+1; GO T
1006	O L; END
1007	C=B+B*B;
1008	CALL ZZZ
1009	(B) ; A[B
100A	]=C; PRO
100B	CEDURE Z
100C	ZZ(M); G
100D	LOBAL C;
100E	C=M+M; E
100F	ND

Figure 9.3 Source-program example as stored in memory.

THE OBJECT-CODE STRING

```
3000   00 XXXXXX    no-op
       90 002220    block
3001   D0 002222    push B
       00 XXXXXX    no-op
3002   F5 34F6XX    push literal |4|
       XX XXXXF6
3003   DF XXXXXX    store
       BB XXXXXX    end-of-statement
3004   D9 001000    load source pointer
       D0 002224    push I
3005   F5 31F6XX    push literal |1|
       XX XXXXF6
3006   DF XXXXXX    store
       BB XXXXXX    end-of-statement
3007   D9 001001    load source pointer
       00 XXXXXX    no-op
3008   00 XXXXXX    no-op
       90 002220    block
3009   D0 002224    push I
       D0 002222    push B
300A   9A XXXXXX    LTE
       B5 003067    branch false
300B   D0 002270    push A
       D0 002224    push I
300C   DD XXXXXX    subscript
       00 XXXXXX    no-op
300D   F5 31F6XX    push literal |1|
       XX XXXXF6
300E   DF XXXXXX    store
       BB XXXXXX    end-of-statement
```

Figure 9.4   First part of the example object program.

The first instruction is a BLOCK instruction, but a BLOCK instruction cannot reside on a word boundary; thus it is preceded by a no-op instruction. The BLOCK instruction points to the block control word at name table location 2220.

The next five instructions correspond to the statement B=4. Notice that the PUSH-LITERAL instruction has the format of a scalar string (the opcode F5 is the string-start character). Since this instruction must start on a word boundary, it is preceded by a no-op instruction. The STORE instruction (address 3003) allocates space for the value currently residing on the top of the stack, updates the identifier control word pointed to by the next item on the stack so that it points to the allocated storage, and deletes the top item on the stack. In this case, no storage would be allocated, since the value (4) is small enough to be stored directly in the identifier control word at location 2222. The END-OF-STATEMENT instruction has two purposes: it causes any operations left pending to occur, and it deletes everything in the stack. The LOAD-SOURCE-POINTER instruction loads the address of the end of the corresponding source statement into a register (for debugging purposes).

The instructions starting at the second halfword of location 3004 through 3007 are quite similar and are associated with the statement I=1.

The next BLOCK instruction is generated because of the label L. A BLOCK instruction is necessary at the point of each label so that the machine can place itself in the proper environment if a GO TO statement in a different block (an inner block) branches to the label. Notice that L's identifier control word in the name table points to location 3008.

The next three instructions evaluate the IF condition, leaving a boolean value on the top of the stack. If its value is false, control transfers to the code for the ELSE alternative (which is absent in this SPL program, so control transfers to the code for the following statement).

The instructions starting at location 300B correspond to the statement A[I]=1, which involves a subscript reference to a structure. The SUBSCRIPT instruction converts the top stack item to an integer value and then searches through the stack for the first noninteger (which would be the reference to the structure) and replaces this item and the upper items [the subscript value(s)] with the address of the designated element. (If multiple subscripts were used, an INTEGERIZE instruction would be generated after the PUSH instruction for all but the last subscript to convert each to an integer.) Since A[I] does not exist at this point, and since the processor does not yet know how much space to allocate for the element, this operation (subscripting) is left pending until the STORE instruction in 300E is executed.

The object code for the statement I=I+1 consists of the eight instructions starting with the second halfword in location 300F in Figure 9.5. The ADD instruction checks for numeric operands and raises an error if one or more is not numeric. The no-op in 3063 is necessary because LOAD-SOURCE-POINTER instructions must reside on word boundaries. The GOTO instruction at location 3065 corresponds to the GO TO statement in the source program. At this point the address of the identifier control word for L is on the stack; the identifier control word points to 3008.

The instructions at locations 3067 through 306A correspond to the statement C=B+B*B and are straightforward. The no-op was placed in 3066 because the PUSH C instruction is the target of a branch (the branch around the ELSE alternative of the IF statement). All the arithmetic instructions check for empirical operands and for a result that exceeds the limit-register value, but neither are applicable in this situation.

THE OBJECT-CODE STRING

The last three instructions in Figure 9.5 implement the CALL statement. The address of the procedure-type identifier control word is pushed onto the stack, as is the address of the argument. When the END-OF-STATEMENT instruction is executed, it sees that an argument exists on the stack; therefore it searches through the stack for a procedure identifier control word. Control is now diverted to location 3072 (the value in the procedure identifier control word).

When the BLOCK instruction at 3072 (Figure 9.6) is executed, a change of environment is detected because the BLOCK instruction points to other than the current name table. Because of this, the machine stores status information in the stack, obtains storage for a new stack, places a pointer to the previous stack in the new stack, and searches the new name table for parameters. Since the argument-transmission method is by name, and since arguments may be expressions with operators, parameters cannot usually be evaluated at procedure-call time. Instead, the parameters are usually initialized to point to one or more object code instructions that are executed whenever the parameter is referenced. However, in this case the argument is a variable; thus the parameter's identifier control word is al-

```
300F   D9 001004      load source pointer
       D0 002224      push I
3060   D0 002224      push I
       00 XXXXXX      no-op
3061   F5 31F6XX      push literal |1|
       XX XXXXF6
3062   AB XXXXXX      add
       DF XXXXXX      store
3063   BB XXXXXX      end-of-statement
       00 XXXXXX      no-op
3064   D9 001005      load source pointer
       D0 002226      push L
3065   95 XXXXXX      goto
       BB XXXXXX      end-of-statement
3066   D9 001006      load source pointer
       00 XXXXXX      no-op
3067   D0 002272      push C
       D0 002222      push B
3068   D0 002222      push B
       D0 002222      push B
3069   AA XXXXXX      multiply
       AB XXXXXX      add
306A   DF XXXXXX      store
       BB XXXXXX      end-of-statement
306B   D9 001007      load source pointer
       D0 002273      push ZZZ
306C   D4 002222      push direct parameter
       BB XXXXXX      end-of-statement
```

Figure 9.5 Continuation of the example object program.

```
306D  D9 001009   load source pointer
      D0 002270   push A
306E  D0 002222   push B
      DD XXXXXX   subscript
306F  D0 002272   push C
      DF XXXXXX   store
3070  BB XXXXXX   end-of-statement
      00 XXXXXX   no-op
3071  D9 00100A   load source pointer
      D7 003078   branch
3072  00 XXXXXX   no-op
      90 002288   block
3073  D9 00100C   load source pointer
      D0 00228C   push C
3074  D0 00228A   push M
      D0 00228A   push M
3075  AB XXXXXX   add
      DF XXXXXX   store
3076  BB XXXXXX   end-of-statement
      00 XXXXXX   no-op
3077  D9 00100E   load source pointer
      B7 XXXXXX   end
3078  B7 XXXXXX   end
```

Figure 9.6 Remainder of the example object program.

tered to point to the argument's identifier control word. An example of the transmission-by-name mechanism for argument expressions is discussed in the next section.

When the PUSH C instruction at location 3073 is executed, the machine recognizes that C is global; therefore it searches backward through the chain of identifier control words pointed to by C's identifier control word until it encounters one that is not global. It pushes this identifier control word onto the stack. When the PUSH instructions at 3074 are executed, the machine sees that their identifier control words are parameters; it pushes the address of the corresponding argument's identifier control word onto the stack (or, in some cases, evaluates the value of the argument). The stack now contains two pointers to the identifier control word for B and a pointer to the identifier control word for C (C in the outer procedure).

The END instruction at location 3077 causes the procedure to terminate. The current stack is deleted, addresses to local variables in the current name table are zeroed, the saved status information in the previous stack is restored, and control is returned to location 306D. The instructions at 306D–3070 performs the assignment to A[B], control branches around the code for the inner procedure, and the program terminates.

Table 9.4 compares this object program with object programs for

THE OBJECT-CODE STRING

equivalent source programs on the Student-PL machine and the S/370. The stored version of the SPL source program was excluded from the SYMBOL numbers to make the comparisons as accurate as possible. SYMBOL appears to be less efficient than the Student-PL machine because of the inefficiencies in the design of its instruction set; for example, 8-bit instructions must be padded with 24 "don't-care" bits, and the instruction-alignment conventions require the insertion of no-op instructions.

Argument Transmission

Since SPL transmits arguments by name, parameters usually cannot be initialized at procedure-call time; instead, the argument must be evaluated whenever the parameter is referenced. As an example, the SPL statement

CALL COST (I+K,20*M)

would yield the object code in Figure 9.7, where the Zs indicate references into some unspecified name table.

The INDIRECT-PARAMETER instructions cause the parameters in the name table for block COST to be initialized to point not to identifier control words for the arguments, but to object-code strings representing the argument expressions. When such a parameter is referenced in the called procedure, the referenced object-code string is executed until a PARAMETER-RETURN instruction is encountered. For instance, whenever a PUSH instruction for the first parameter in COST is executed, the instructions PUSH I, PUSH K, and ADD are executed in its place.

Table 9.4 Relative Program Sizes

	Instruction Executed	Object-Code Size (bytes)	Total Program Size (bytes)
SYMBOL	81	324	396
Student-PL	62	96	369
S/370	303[a]	564	5912
S/370[b]	112[a]	360	5496

[a] Plus approximately 90,000 instructions at the beginning of the program to set up storage for the call/return mechanism.

[b] The PL/I version of the program without the ALLOCATE statement and without subscript checking.

```
1000   D0 ZZZZZZ    push COST
       D7 001006    transfer
1001   D0 ZZZZZZ    push I
       D0 ZZZZZZ    push K
1002   AB XXXXXX    add
       D8 XXXXXX    parameter return
1003   F5 3230F6    push literal |20|
       XX XXXXF6
1004   D0 ZZZZZZ    push M
       AA XXXXXX    multiply
1005   D8 XXXXXX    parameter return
       00 XXXXXX    no-op
1006   D5 001003    indirect parameter
       D5 001001    indirect parameter
1007   BB XXXXXX    end-of-statement
```

Figure 9.7 Transmission-by-name object code.

EXERCISES

9.1 One internal form of a scalar is the numeric form. This has not yet been discussed, but guess what its representation is likely to be.

9.2 What is the linear representation of the structure <0000000 | 1111111>?

9.3 Describe the contents of the identifier control word in location 2224 after the instructions in location 3006 in Figure 9.4 have been executed.

9.4 Describe the contents of the identifier control word in location 2224 if the value 1234.5678 was assigned to I.

9.5 In Figure 9.4, why are there two consecutive no-op instructions in words 3007 and 3008? Why couldn't both be eliminated, placing the BLOCK instruction in the second halfword in location 3007?

9.6 When the object code corresponding to the statement C=M+M in the example is executed, from what identifier control word is the value of M obtained? What identifier control word is altered by this statement?

10 | SYMBOL Processor and Configuration Architecture

This chapter purposely deviates from the main line of thought of the book (i.e., computer architecture as defined in Chapter 1) in that it discusses the implementation of SYMBOL's processors and how they interact with one another. The reasons for this are twofold. First, the computer architecture of SYMBOL (the computer architecture of the central processor) has many implications on the implementation of the system (e.g., how memory is managed), and it is worthwhile to examine these implications. Second, one of the unique aspects of SYMBOL is its distributed-processing, dedicated-function approach to multiprocessing; thus a more thorough analysis is warranted.

THE MAIN BUS

Since the main bus is the communication path among all the processors (Figure 8.1), understanding its use is the first step in understanding the internal operation of SYMBOL. The main bus is used for

1. Transmission of a memory request from a processor to the memory controller (memory request cycle).
2. Transmission of the results of a memory request from the memory controller to a processor (memory output cycle).
3. Transmission of a control signal from a processor to the system supervisor (control exchange cycle).

4. Transmission of a control signal from the system supervisor to a processor (control exchange cycle).
5. Transmission of data from a processor in the central processor to another processor in the central processor (sneak cycle).

The main bus consists of 111 lines. The meaning of the lines is illustrated in Table 10.1. The bus is used on a priority basis; lines MP1–MP8 are priority lines (MP1 being the highest priority and MP8 the lowest). When a processor must use the bus, it raises its priority line. For example, if the central processor wants to communicate with the memory controller, it raises MP3; if it wants to communicate with the system supervisor, it raises MP6. If none of the higher priority lines is raised, the processor uses the bus during the next cycle (clock period). If a higher priority line is raised, the processor waits for one cycle and retries the request.

As an example, assume that the central processor desires to send a word of data to memory. It would raise (set) MP3 and test MP1, MP2, and MP7. If MP1 and MP2 are not set and if MP7 is set (indicating that the memory controller is not processing another request), the central processor would place the memory address on lines MA, the data on lines MD, a code indicating the desired memory controller operation on lines MC, and an identification of the terminal or user associated

Table 10.1 The Main Bus

Line	Name	Use
1		Not used
2-7	MC	Memory controller function and completion codes
8-12	MT	Terminal number
13-36	MA	Memory Address
37-100	MD	Data
101		System clear
102		System clock
103	MP1	System supervisor memory request
104	MP2	Interface processor memory request
105	MP3	Central processor memory request
106	MP4	Translator memory request
107	MP5	Indicates next cycle is a memory output cycle
108	MP6	Control exchange cycle request
109	MP7	Indicates memory controller is free
110	MP8	Memory reclaimer memory request
111		Not used

with this request on lines MT. The memory controller now drops line MP7 and proceeds to process the request. During this time the bus may be used for processor-to-processor communication but not for memory requests. When the memory controller is finished, it raises MP5. In the next cycle the memory controller places a completion code on lines MC, and, if it is returning an address (which is usually the case because of the list-structure concept of memory), it places the address on lines MA. During this time the central processor has been testing line MP5. When it sees the line raised, it knows that the memory controller's results will be on the bus during the next cycle.

If a processor wishes to send a control signal to the system supervisor or if the system supervisor wishes to send control signals to one or more processors, it raises MP6, and, if MP1-MP5 are not set, it places its control information in MD. Such usage of the bus is called a control exchange cycle, and during such a cycle all processors may potentially use the bus. This was accomplished by partitioning the MD lines into distinct groups during a control exchange cycle. For instance, during a control exchange cycle, lines 37-40 are signals from the system supervisor to the central processor, lines 41-44 are signals from the central processor to the system supervisor, lines 45-48 are signals from the system supervisor to the translator, lines 49-52 are signals from the translator to the system supervisor, and so on.

As mentioned earlier, the MC lines are used during a memory request cycle to communicate a function code to the memory controller. Lines 2-5 designate which of 14 possible logical-memory requests is to be performed, and lines 6-7 indicate with which of the user's virtual-memory page lists the request is associated (the latter is used to minimize storage fragmentation; both are explained in the next section). During a memory output cycle, the memory controller places a completion code on the MC lines. For instance, the value 000011 indicates successful completion of the request, 000001 indicates unsuccessful completion because of a page fault, and 100000 indicates a memory hardware error.

MEMORY MANAGEMENT

The high-level, flexible, and dynamic memory-management mechanisms are perhaps the most significant aspects of the SYMBOL system. As mentioned in Chapter 8, these mechanisms are achieved by the memory controller and, to a lesser degree, the memory reclaimer and the system supervisor.

SYMBOL contains a hierarchy of storage concepts. At the top of the hierarchy is logical storage. This view of storage is a limitless supply of limitless-length linear lists or strings of 64-bit words. Next in the hierarchy is virtual storage. This view of storage is a contiguous vector of 16,777,216 64-bit words. The job of the memory controller is to provide the abstraction of logical storage by mapping it into virtual storage. Virtual storage is partitioned into pages containing 256 words (2048 bytes or characters). One of the functions of the system supervisor is to map virtual-storage pages into the third storage concept: main memory.

Logical Storage

Logical storage is the view of storage presented to the other processors by the memory controller. The best way to define logical storage is to define the 14 memory functions provided by the memory controller; they are listed in Table 10.2.

Chapter 9 pointed out that entities such as the name table and object-code string are not necessarily stored in consecutive virtual-storage locations and that this is transparent to all but the memory controller. The use of the operations in Table 10.2 should now substantiate this. For instance, when the translator is building a name table, it sends an assign-group request to the memory controller. The memory controller returns the address of the first word in this new string. When the translator is ready to store the block control word (the first word), it requests a store-and-assign operation and passes the memory controller the address obtained as a result of the assign-group operation and the value to be stored. The memory controller returns the address of the next location in the string. To complete the name table, the translator sends store-and-assign requests, passing as an address the address returned from the previous store and assign.

As another example, the central processor sends fetch-and-follow requests to the memory controller to fetch the machine instructions. For each request, the memory controller returns a 64-bit word (normally two instructions) and the address of the next instruction. If the current instruction has an operand address, the central processor would use the operand address in a fetch-and-follow request to obtain the identifier control word.

Since the addresses passed to the memory controller are addresses that were previously assigned and returned by the memory controller, the fact that words in a logical-storage string may not be in consecutive locations in virtual storage is transparent to the other processors. If fact, the processors do not view these addresses as traditional addresses; to the processors, the "address" is simply a 24-bit unique name of some value in storage.

Table 10.2 Memory Controller Operations

Operation	Purpose and Function
AG Assign Group	Used to begin a new string. Returns the address of the first word in the string.
SA Store and Assign	Used to store a value in a string. Stores a specified value at a specified address and returns the address of the next word in the string.
FF Fetch and Follow	Used to scan a string. Returns the value at a specified address and returns the address of the next word in the string.
FR Fetch and Reverse Follow	Used to scan a string backwards. Returns the value at a specified address and returns the address of the preceeding word in the string.
FL Follow and Fetch	Used to scan a string. Returns the value and address of the word beyond a specified word in the string.
IG Insert Group	Used to insert a string within another string.
SI Store and Insert	Used to store a value in a string and insert another string at this point.
SO Store Only	Used to store a value in a string. Stores a specified value at a specified address.
DS Delete String	Used to delete a string.
DE Delete to End	Used to delete part of a string. Deletes all words in a string beyond a specified address.
FD Fetch Direct	Used to fetch the value of a word at a specified address in main (real) memory.
SD Store Direct	Used to store a value at a specified address in main (real) memory.
DS Delete Page List	Used by the memory reclaimer to add a list of virtual-storage pages to the free-page list.
RG Reclaim Group	Used by the memory reclaimer to move a group of words from the garbage list to the available-group list.

Virtual Storage

The function of the memory controller is to map this concept of logical storage into a single virtual address space of 16 million words. To understand how this rather complicated process is performed, we can first examine the overall definition of virtual storage as illustrated in Figure 10.1.

The first page of virtual storage is used for system tables. The next three pages contain tables of information about each user. The next two pages contain the input/output buffers used by the channel controller. The remaining pages are the address space from which the memory controller creates the logical-storage concept. As indicated in Figure 10.1, because of the logical-storage concept there is no need to give each user a contiguous area of virtual storage. In fact, the pages for different users are usually interspersed in virtual storage, although storage for different users is never allocated from the same page.

The format of a page in this last area is shown in Figure 10.2. The first four words represent the page header; the header is used primarily

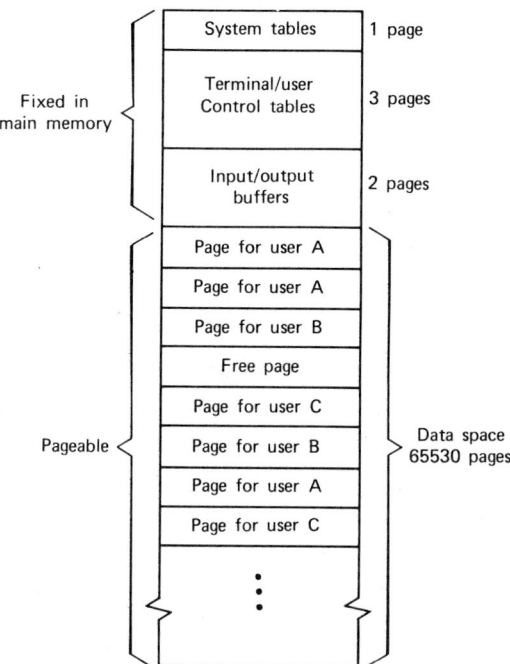

Figure 10.1 Virtual-storage page allocations.

MEMORY MANAGEMENT

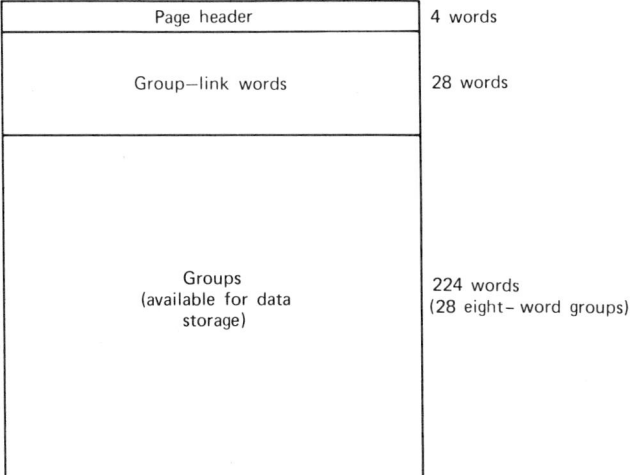

Figure 10.2 Format of storage within a page.

to link pages together on various page lists. The next 28 words are *group-link words*. Their purpose is to create the logical-storage view of memory. The last 224 words represent the space on each page that is usable for data storage.

The 224 words of usable storage on each page are subdivided into 28 eight-word allocatable groups of storage, and each 8-word group is associated with a group-link word. Turning back to some of the examples in Chapter 9, we see that although a string such as a name table or object-code string is not necessarily stored in consecutive virtual-storage locations, each eight-word division of the string is stored in consecutive locations. The reason for this should now be apparent.

To understand how the concept of logical storage is created, examine the format of a group-link word in Table 10.3. Each group-link word has forward and backward pointers; these are virtual addresses of the group-link words associated with the next and previous groups in this logical-storage string. That is, logical-storage strings are represented by chains of group-link words and by the groups associated with these group-link words. Note that the addresses in Table 10.3 are 24-bit virtual addresses. This implies that group-link words can point to group-link words on other pages; therefore, a logical-storage string can span pages.

To understand how storage is allocated, assume that some processor that is processing on behalf of user A sends an assign-group request to the memory controller. The memory controller first searches for a page

Table 10.3 The Group-link Word

Bits	Meaning
0–7	Rapid search field
8–31	Address of next group-link word in string
32	A rapid search field is present
33	A forward-link field is present
34	A backward-link field is present
35	Space is available in the group
36	This is the last group in a string
37	One of the words in the group is an address
38–39	Unused
40–63	Address of previous group-link word in string

of virtual memory associated with user A that contains some available space. (We ignore, for now, the process used by the memory controller to find such a page. Also note that if one cannot be found, the memory controller would find an unused page and assign it to user A.)

Assume that a page is found, and, on this page, groups 1, 3, and 5 are allocated, and the remaining groups are available. This information is represented in the following way. The page header contains two 5-bit fields called available group list start (AGLS) and initial group assignment counter (IGAC). In this situation AGLS would contain the value 2, IGAC would contain the value 6, group-link-word 2 would contain the virtual address of group-link-word 4, and bit 33 of group-link-word 4 would be off. In other words, available groups are placed in a list, and AGLS contains the number of the first group in the list. However, rather than searching this list initially, if the last n groups in a page are available, they are not linked together; instead IGAC contains the number of the first one.

To perform the assign-group operation, the memory controller sees that IGAC is not 29; it computes the virtual address of the first word of group 6, adds one to IGAC, and returns this address to the requesting processor. If IGAC had been 29, the memory controller would have taken the group from the available-group list.

If the next request is a store and assign, the memory controller must return the address of the next word in the string. Seeing that the address given to it is not for the last word in a group, the memory controller simply increments the address by one and returns it. After seven more store-and-assign operations for this string, the memory controller sees that it is storing into the last word of a group. At this point it goes through the process of finding an available group as described earlier.

MEMORY MANAGEMENT

Assuming that it finds that group 7 is available, it places the address of group-link-word 7 in the forward-link field of group-link-word 6, places the address of group-link-word 6 in the backward-link field of group-link-word 7, and returns the address of the first word in group 7. In general, then, the memory controller services seven out of eight requests rapidly, but the eighth request involves considerably more overhead.

The operation of the memory controller for other operations should be apparent. For instance, for fetch-and-follow operations, the memory controller returns the specified address plus one if the word being fetched is not the last word in a group. If the word is the last word in a group, the memory controller must find the next address by going to the associated group-link word, finding the next group-link word by following the forward link, and returning the address of the first word in this group.

Group Reclamation

When a processor has no further use for a string, it sends a delete-string request to the memory controller. The process of freeing the storage occupied by a string is a potentially time-consuming operation. The end result is to add the groups occupied by the string to the available-group list, but this is not as easy as it might first appear. Since a string may span pages (but each page has its own available-group list), the string must be broken apart by page, and the groups of the substrings must then be added to the available-group lists for each page. In addition to being time consuming in itself, this process could generate page faults, since some of the pages might not currently reside in main memory.

Another complication is that a string might point to other strings, and the deletion of a string implies that all associated strings must be deleted. An example of this is an SPL structure. Assume that A is a structure and that the string representing A is to be deleted. If some of the elements of A are structures, the string representing A contains addresses of other strings, and these strings may in turn point to other strings. For instance, if A[1] is a structure, the first word in the first group of the string representing A contains an address of another string rather than a data value.

If these steps were performed by the memory controller whenever it received a delete-string request, the memory controller would be busy for an inordinate amount of time, bringing the entire system to a virtual standstill. The separate memory-reclaimer processor is the solution to this problem. The memory controller processes a delete-string request rapidly by simply adding the string to one of several garbage lists, leaving the reclamation of the space up to the memory reclaimer.

The memory reclaimer has the lowest-priority access to memory; it may only send requests to the memory controller when the memory controller is free, when no other processors are about to send requests to the memory controller, and when a control exchange cycle is not in progress. The memory reclaimer continuously scans the garbage lists. When it finds a nonempty list, it removes the first string and scans all the group-link words in the string. (The memory controller accomplishes this by sending the standard requests to the memory controller.) The memory reclaimer adds each group to the available-group list on its corresponding page by sending the reclaim-group request to the memory controller. Thus the memory controller has a role in memory reclamation, albeit a passive one, by acting as a "subroutine" of the memory reclaimer.

If there is any possibility that the string contains addresses of other strings, the memory reclaimer scans each word in each group of the string, looking for addresses rather than data values. If it finds an address, the memory reclaimer adds it to the garbage list; thus such strings will be reclaimed during subsequent passes through the garbage list.

Page Allocation

The discussion of space allocation was purposely vague on how the memory controller finds a page with available space. In this section the mechanism is clarified, and the management of the pages of virtual storage is discussed.

Virtual-storage pages are managed in a way analogous to the management of groups, that is, by the use of available and garbage lists. One of the system tables in main memory contains a header (initial address) of an available-page list. All pages that are not currently associated with a user appear in this list. The header area in every page contains a 24-bit field used to link the page on a page list. In the case of available pages, this field links the page in the available-page list.

As mentioned earlier, a control table for each user exists in main memory. All the pages currently associated with a user are linked together, but rather than using a single list, each user has three page lists (TPL1, TPL2, and TPL3). Headers for these three lists exist in each control table.

The purpose of the three page lists is to minimize the number of

MEMORY MANAGEMENT

pages residing in main memory by allocating space for items that are likely to be used together on the same page(s). The system convention is that TPL1 contains the pages containing the user's transient working area (e.g., his source program), TPL2 contains the pages containing the user's name tables and space for the program's variables during execution, and TPL3 contains the pages containing the user's object-code string.

In addition to being on one of the user/terminal page lists, a page may also be on an available-space list if the page contains one or more available groups. Associated with each TPL header in the user control table is a header for an available-space list. A separate field in the page header is used as a link field for this list.

Figure 10.3 illustrates these lists. To avoid overcomplicating the diagram, only nine pages are illustrated, and only one user control table is shown. The diagram is not as complicated as it appears if one follows each list separately. Pages 2, 4, and 9 are currently not allocated; they appear on the available-page list. Page 3 is on user A's TPL1, and, since it contains some free space, it is also on the TPL1

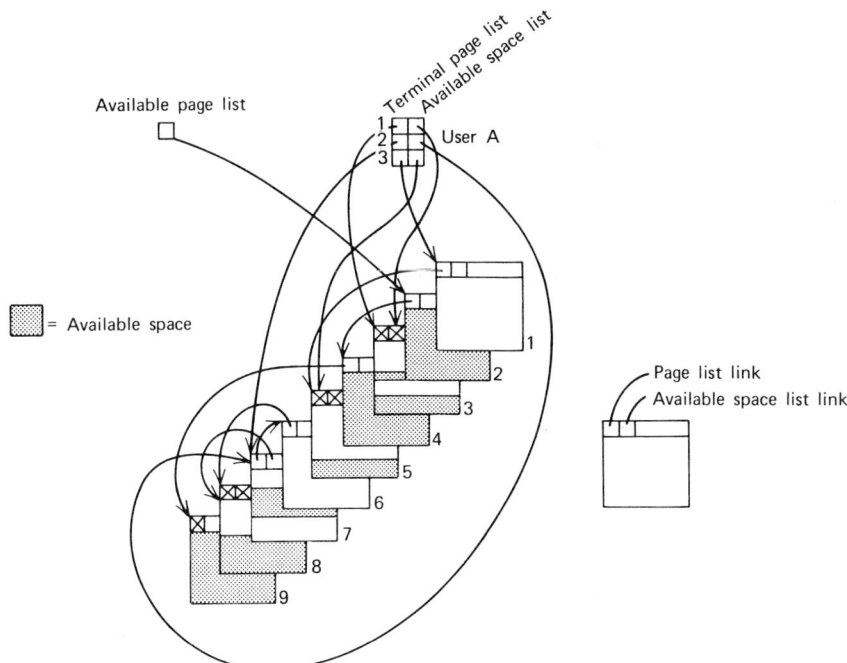

Figure 10.3 Example of SYMBOL's page lists.

available-space list. Pages 7, 6, and 8 are on TPL2; pages 7 and 8 contain some available space; they are on the available-space list associated with TPL2. Pages 1 and 5 are on TPL3, and page 5 is on the associated available-space list.

As discussed earlier, when a processor has a memory request, it places a function code on the six MC lines of the main bus. The processor places a code on four lines, indicating which of the 14 operations is to be performed. On the remaining two lines it places a code indicating to which of the three user page lists the request is associated (this code is only used for the assign-group operation).

Assume that the central processor sends the memory controller an assign-group request and indicates that it should be associated with TPL2. The memory controller examines the available-space list associated with TPL2. If it is not empty, it assigns the group from the first page in the list. If the available-space list is empty, the memory controller removes a page from the available-page list, adds it to TPL2 and TPL2's available-space list, sets the page's IGAC (initial group assignment center) to 1, and assigns the group from this page.

Page Reclamation

Page reclamation is performed by the memory reclaimer in a process similar to group reclamation. Assume that a user has just compiled and executed his program and that he now wants to run another program. When the system supervisor receives the RUN command, it puts the user's TPL2 and TPL3 on the memory reclaimer's queue and zeroes the user's TPL2 and TPL3 headers. The memory reclaimer eventually adds each page to the available-page list by sending a delete-page-list request to the memory controller.

When a page is reclaimed, the memory reclaimer clears the page to zeroes so that group allocation and reclamation will not become confused during later use of the page. It is not necessary to delete each string on a list of pages being reclaimed; thus the number of delete-string requests processed by the memory controller is less than the number of assign-group requests.

THE SYSTEM SUPERVISOR

The system supervisor is the processor that performs the functions normally associated with an operating system. The other processors can be viewed as slaves to the system supervisor in that the system supervisor tells them which work to perform. On the other hand, the

THE SYSTEM SUPERVISOR

system supervisor is a passive processor; it is interrupt driven by the other processors. It services these interrupts either by giving some work to another processor or by performing some task itself.

The main functions of the system supervisor are managing the other processors by maintaining their work queues, scheduling paging operations when a page fault occurs, and managing the system and user control tables. The reason that the system supervisor, rather than the memory controller, controls the paging process is that the system supervisor has a global view of the system's operation, allowing it to make more intelligent page-replacement decisions.

The Paging Process

The system supervisor continually monitors the MC lines during a memory-output cycle on the main bus. If the MC lines contain the value 000001 (page fault), the system supervisor makes a note of the address currently on the bus; thus it knows which virtual-storage page must be fetched into main memory. If the system supervisor sees the completion code 100011, which indicates a successful completion of a memory request but the memory controller had to remove a page from the available-page list, the system supervisor increments a page-usage counter associated with the user. The page-usage counter is used to terminate users who exceed some threshold (e.g., a user whose program is in a run-away loop that is allocating excessive amounts of storage).

The paging process begins when a processor sends the system supervisor a control signal indicating that it has encountered a page fault. The system supervisor executes a page-replacement algorithm (a hardware circuit, since the system supervisor contains no software), and it places paging requests on the queue for the disk controller (possibly a request to move a main-memory page to the paging disk and a request to read a paging-disk page into this main-memory page). The system supervisor then finds the user's entry on the queue for the faulting processor and marks the entry as being in page-wait state (but it leaves it on the queue in its current position). The page-faulting processor is then assigned another queue entry to process.

The paging disk is divided into four quadrants; each page resides on one quadrant, and each virtual-storage page has a distinct reserved position on the disk. As the disk turns, it notifies the disk controller of which quadrant is about to appear next under the read/write heads. The disk controller passes this information on to the system supervisor. The system supervisor uses this information to minimize rotational delay, that is, it uses this information to direct the disk controller to the appropriate entry in its queue.

There is another type of page list, not illustrated in Figure 10.3. This list is the in-core list, which is a list of pages that currently reside in main memory. Another field in the header of each page is used as a link field for the in-core list. When a paging operation has been completed, the system supervisor puts the new page at the bottom of the in-core list. (The "paged-out" page was removed from this list when the paging requests were created.) The system supervisor then completes the process by removing the page-wait mark in the user's queue entry associated with the faulting process.

Page Replacement

The page-replacement algorithm of the system supervisor is interesting because it is a FIFO (first in, first out) algorithm that gives preference to the program at the top of the central processor's queue and uses global system information to determine the optimal page to remove from main memory.

The system supervisor computes a *page-residency index* for each page in main memory, which is a measure of the system supervisor's desire to retain the page in memory. The index is assigned according to the rules in Table 10.4. The pages that have an index of 0 are primarily pages that have a low probability of usage in the immediate future; these include free pages, pages belonging to a user who is in a pause state or whose program has just completed execution, and pages associated with a user who has just exceeded his alloted time slice on a processor.

As part of the page-replacement algorithm, the system supervisor also computes a *page-push priority* of the user that encountered the page fault. If the user is at the top of the central processor's queue, the priority 3 is assigned. If the user is at the top of the translator's or interface processor's queue, the greater of the following two priorities is assigned: 2, or a priority associated with the user (with possible values 0, 1, 2, or 3). For a user with any other status, the priority associated with the user is assigned.

To find a page to be removed from main memory, the system supervisor scans the in-core list from top to bottom. As soon as a page is found whose page-residency index is less than the page-push priority, this page is selected. If no such page is found, the first page whose page-residency index equals the faulting user's page-push priority is selected. If no eligible page is found, the system supervisor gives up

THE SYSTEM SUPERVISOR

Table 10.4 Assignment of the Page-residency Index

Page-residency Index	Pages
3	Pages belonging to the user at the top of the central processor's queue
2	Pages belonging to other users in the central processor's queue
2	Name-table pages (i.e., pages in TPL2) belonging to users in the translator's queue
1	Pages belonging to users in the interface processor's queue
1	Non-name-table pages belonging to users in the translator's queue
0	All other pages

and does not mark the user's queue entry as page waiting. Eventually the faulting processor will encounter the same page fault; hopefully at this time the dynamics of the sysem will have changed so that the page fault can be resolved.

As implied by this discussion, the page-replacement algorithm does not blindly select a page to be removed; it uses information about the status of the user generating the fault, the contents of each page, and the status of the users associated with each main-memory page to make an intelligent decision. The user in the top position in the central processor's queue receives the best service. He can steal pages from anyone else (including his own pages as a last resort), but no other user can steal main-memory pages associated with him.

Processor Queue Management

The system supervisor is also responsible for adding entries to each processor's queue, deleting entries, and informing each processor of the queue entry that it is to process. The algorithms are logic circuits in the system supervisor, but they are influenced by parameters that are stored in main memory. The queues are maintained in the system-table area of main memory.

When a processor is available to perform some new work, the system supervisor specifies the queue entry to be processed by scanning the processor's queue from top to bottom. The entry that is selected is the first entry that is not in a page-wait state. When an entry is added to a queue, it receives a *queue run time* that is initialized from a system parameter. At periodic intervals, the system supervisor decrements the queue-run-time values of all entries that are currently receiving service from a processor. If an entry's queue run time reaches zero, the entry is moved to the bottom of the processor's queue, and its queue run time is reinitialized. Thus a user receives good service from a processor until his queue run time expires, at which time his probability of receiving service drops. Also, new entries are always added to the top of the queues, ensuring that they (at least initially) receive priority service.

Another parameter controls the depth to which the system supervisor searches a processor's queue. If the queue is larger than this value, users who exceed their time slice and move to the bottom of the queue temporarily fall outside the processor's multiprogramming set.

The queue run time serves to control processor-bound tasks, but it does not prevent a paging-bound task from monopolizing main memory and the processor, since such a task would tend to sit at the top of a queue for an inordinate amount of time. Since this situation is most likely to occur on the central processor, the top entry on the central processor's queue is given a *top-time* parameter. If the entry sits on the top of the queue for longer than this time, it is demoted to the bottom of the queue.

System Software

As mentioned earlier, SYMBOL contains virtually no system software. In fact, the system does contain a small amount of system software, but the software only performs a few auxiliary functions, and the system can operate without it.

The system software is used to report certain errors to the terminal user (e.g., syntax and execution errors), to perform certain file-management functions, to perform accounting functions at the beginning and end of a terminal session, and to provide an on-line program-debugging package. The system software is written in a slightly extended version of SPL and is stored in virtual memory as SPL object programs. When the system supervisor wishes to invoke a software function, it creates an entry for the function on the central processor's queue. Thus the system software is executed by the central processor, and it competes with the other user programs for the system's resources.

System programs are written in a restricted version of SPL. The privilege to be compiled with the restricted language statements is the only distinction between a "privileged" (system) program and a "regu-

THE CENTRAL PROCESSOR 141

lar" program. The majority of the restricted statements are statements that give a program direct access to the 14 operations of the memory controller. Examples of the restricted statements are assign group, fetch and follow, store and assign, and so on.

THE CENTRAL PROCESSOR

The function of the SYMBOL central processor is to execute SPL object programs. In doing so it deals with four types of storage: the object program, the name tables, the stack, and data storage. The central processor's stack is a list in virtual storage; the central processor uses the normal memory-controller operations to access the stack.

Part of each user's control table in the low area of main memory is used by the central processor to keep track of the state of the user's program. Among other things, the central processor maintains in these control tables the address of the object-code string, the address of the current name table, the address of the current stack, the address of the current machine instruction, and the current LIMIT value.

Internal Data Representations

Chapter 9 implied that the central processor maintains data in a more efficient internal representation, although it converts data to and from its external representation when communicating with another processor (i.e., the translator and interface processor). The external representation of a scalar, as illustrated in Chapter 9, is a string. If the scalar does not have a numeric value, its internal representation is identical to its external representation. If the scalar has a numeric value, it is stored in an internal form called the *numeric* form.

The numeric form is a packed-decimal (two digits per character) mantissa/exponent (floating-point) representation. Numerics are stored in an integral number of words; the high-order bit in the last character of the last word must be on. The first character is a control character designating the value as numeric and specifying the exponent and mantissa signs. F0 designates a positive mantissa and exponent, F1 designates a negative mantissa and positive exponent, F2 designates a positive mantissa and negative exponent, and F3 designates a negative mantissa and exponent.

The second character contains the value of the exponent (00 to 99). The first four bits after the last mantissa digit must be F or E; F designates an exact value, and E designates an empirical value. For example, the value 1283 is represented as

FO 04 12 83 FX XX FX

and the value $-1234567.891234567EM$ is represented as

F1 07 12 34 56 78 91 00
00 00 23 45 67 EX XX EX

(As indicated above, only the third through seventh characters in each word are used for the mantissa.)

The internal representation of structures is called the *normal representation*. In this form, structures are stored as one or more vectors, where a vector is represented by a list (string) in storage. The components of a vector are stored consecutively in the list; the list is terminated by a word beginning with the F7 (end-of-vector) control character. However, if a component is a structure, it is represented in the list as a substructure pointer. The format of a substructure pointer is

EC AAAAAA SS BBBBBB

AAAAAA is the address of the vector representing the substructure. SS is the subscript value of the most recently accessed component of the vector, and BBBBBB is the address of the most recently accessed component. (As indicated in Table 9.3, these last two fields are also present in identifier control words for structures.) These last two fields are used to rapidly perform subscripting operations. They are based on the premise that if the Ith component was the component most recently accessed, the next component to be accessed is likely to be the Ith component or one beyond it (e.g., the Ith+1 component). When indexing into a component, then, the central processor compares the subscript to SS. If it is greater than or equal to SS, it can begin the search from address BBBBBB rather than AAAAAA.

Figure 10.4 illustrates the internal representation of a structure being used as an irregular array. The first vector in the structure has three components, two of which are substructure pointers. All scalar values are represented in their numeric internal forms. If we assume that the last reference to this structure (call it S) was S[2,1], and that prior to this a reference was made to S[1,3], the "quick search" fields (subscript values and component addresses) are those illustrated.

The Four Subprocessors

The central processor consists of four subprocessors, each of which has its own set of registers and arithmetic units. The subprocessors are connected to an 8-bit control bus that is used to send control signals

THE CENTRAL PROCESSOR

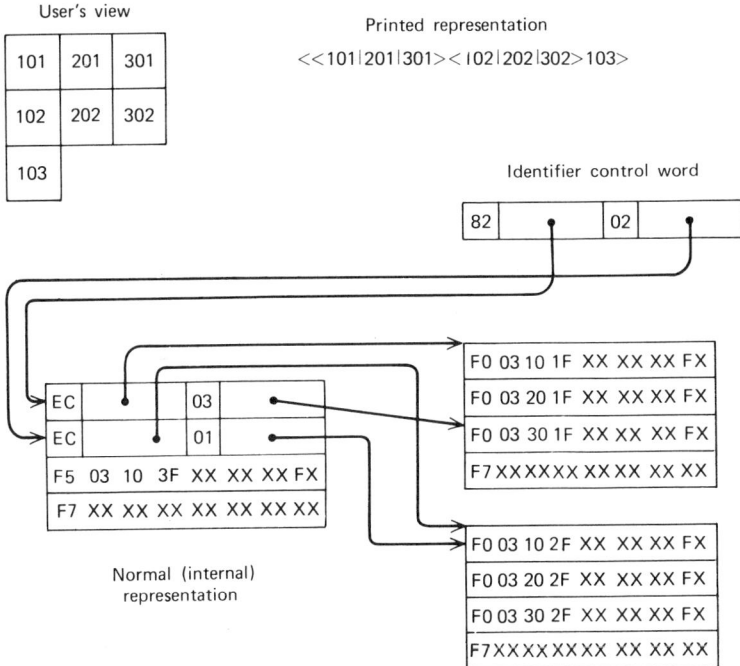

Figure 10.4 Internal representation of a structure.

among the subprocessors. The four subprocessors are also connected to the main bus, which is used to send data among the subprocessors. The subprocessors use the main bus when it is not being used by other processors.

The main subprocessor in the central processor is the instruction sequencer. The instruction sequencer fetches the next machine instruction and processes the literal, branching, and procedure-call instructions. For other instruction types, the instruction sequencer invokes one of the other subprocessors to process the instruction.

The second subprocessor is the arithmetic processor, which handles all arithmetic and numerical-comparison instructions. The precision-controlling processes and the use of the limit register are functions of the arithmetic processor.

The third subprocessor is the format processor, which handles all scalar and boolean instructions and performs automatic data conversions between the string and numeric representations.

The fourth subprocessor is the reference processor. It handles assignment instructions and references to the name table on behalf of the other processors, but its main responsibility is dealing with structures. The reference processor is responsible for converting between the linear and normal representations and handling difficult situations, such as assigning a structure to a structure element and assigning a scalar to a structure element where the size of the scalar is greater than the size of the existing element.

A major part of the reference processor is devoted to structure subscripting operations. It uses the "last used subscript fields" discussed earlier to increase the efficiency of those operations (i.e., to avoid always having to scan from the beginning of the vector). However, subscript operations are potentially time consuming, because elements can be of differing sizes and can span words. To avoid requiring the reference processor to scan every word in a string to count over to the required word, the 8-bit rapid-search field in the group-link word is used (Table 10.3). These 8 bits correspond to the eight words in the group. If a rapid-search bit is on, it indicates that the word is the first word of a vector element (component). Hence if the reference processor wants to locate the 14th component, it scans across the linked group-link words until it encounters the 14th set rapid-search bit, rather than scanning each word of the corresponding groups.

THE TRANSLATOR

The translator is the processor that translates (compiles) SPL programs into an object-code string and one or more name tables. As mentioned in Chapter 8, this process is performed by hardware sequential-logic circuits rather than by software.

The translator is logically divided into two sections: the code-generation section and the name-table section. The code-generation section scans the source program and builds the object-code string. When the scan encounters a nonliteral letter (which could be the start of an identifier or a reserved word), control is given to the name-table section. The name-table section continues the scan until a delimiter is encountered. It then searches a table of reserved words stored in main memory, and, if the name is a reserved word, it returns the op-code to be generated.

If the name is not a reserved word, the current name table is searched. If the name exists, the address of the identifier control word is returned. If not, the name and an identifier control word are appended to the name table, and the address of the new identifier control word is returned.

After the name tables and object-code string have been completed, the "linkage editing" process must be performed. The name-table section scans the name tables, looking for global identifiers. When one is found, the back pointer in the block control word is followed, and the global identifiers are linked to their corresponding identifiers in the outer blocks.

A second linking process is concerned with unresolved procedure identifiers. If one is found, the user and system libraries are searched for the missing procedure. Since programs are stored in libraries in their source forms, the compilation process must be continued, creating new name tables and appending the additional object code to the end of the existing object-code string.

THE REMAINING PROCESSORS

The remaining processors are the interface processor, which transfers data between virtual storage and the I/O buffers, performs the editing operations entered from the terminal, and handles I/O operations initiated by object programs; the channel controller, which moves data between the low-speed I/O devices (e.g., the terminals) and the I/O buffers; the disk controller, which moves data between main memory and the secondary-storage devices; and the maintenance processor. The implementation of these processors is fairly straightforward; no further discussion is warranted.

SIGNIFICANCE OF THE SYMBOL SYSTEM

The architecture and implementation of SYMBOL is significant from many points of view, although SYMBOL is not free of problems. The most significant aspects of the system are its approach to multiprocessing and memory management. SYMBOL's multiprocessor configuration is unique in that, rather than containing multiple generalized "peer" processors, it contains specialized processors that perform distinct parts of the computing system's functions. The memory-management mechanism is unique because of the high-level memory abstractions provided by the memory controller. Higher-level data objects such as stacks, queues, variable-length strings, and list structures that are needed by the other processors are easily created through the use of the logical-storage operations, thus reducing the total amount of memory-management logic needed in the system.

The "all hardware" implementation of the system is significant in some ways and a source of problems in others. The hardware system supervisor and translator contribute significantly to the phenomenal "compile, load, and go" speed of the system and the speed in which page faults are resolved. Also, the logical-storage abstraction would not have been feasible if it had been implemented in software. It also demonstrates the feasibility of embodying traditional software functions "into silicon." However, the "all hardware" implementation also presents some large problems, particularly the large cost of making changes to, or extending the function of, the system. It is likely that the same goals could have been achieved, and some of the problems eliminated, by the use of microprogrammed processors.

In respect to the architecture of the central processor, the large semantic gap present in most systems between the language and the architecture has been largely eliminated. The architecture includes many of the concepts discussed in Chapter 4, including self-describing data, variable-length data, and decimal arithmetic. The notable weakness in the architecture, as mentioned in Chapter 9, is the inefficient design of the instruction set.

The major problem impeding the use of SYMBOL as a general-purpose system is, of course, its support of only a single language, compounded by the fact that the language is new and unfamiliar. One solution to this problem might be the inclusion of several translator processors in the system (e.g., an SPL translator, a FORTRAN translator, a COBOL translator), but this would not prove feasible because of the high SPL orientation of the central processor. A possible direction, then, in addition to providing translators for each language, is to provide multiple central processors, each with an orientation toward a specific high-level language.

EXERCISES

10.1 Why are the MT lines included in the main bus? In other words, why is the terminal/user identification transmitted during a memory request?

10.2 The central processor frequently requests a fetch-and-reverse-follow operation of the memory controller. For what purpose?

10.3 When the memory controller is allocating a new group, its use of the initial group assignment counter (IGAC), where applicable, does not appear to be significantly faster than taking a group from the available-group list. What, then, is the value of IGAC?

10.4 Why are group-link words doubly linked? That is, why does the group-link word contain a backward-link field?

EXERCISES

10.5 Why does the memory controller provide the reclaim-group and delete-page-list operations for use by the memory reclaimer, since the memory reclaimer could perform these functions by using other operations such as store-only?

10.6 Must the memory reclaimer scan every string to determine if it contains addresses to other strings?

10.7 In scanning strings for addresses, how does the memory reclaimer distinguish between addresses and data values?

10.8 What are the differences among the available-page, available-space, and available-group lists?

10.9 How would the normal representation of the structure in Figure 10.4 change a a result of the following assignment statement?

A[2,2] = <1234|4321>

IV
A Multiple-Language-Directed Architecture

11 | The Burroughs B1700 System

When one attempts to design a general-purpose computing system (i.e., a system providing multiple programming languages) with a language-directed architecture, the most significant problem is significantly shrinking the semantic gap between the architecture and a particular programming language, but, as a result, introducing large semantic gaps between the architecture and the other programming languages. This problem is illustrated in Figure 11.1, which is a conceptual graph of the semantic differences among languages and architectures. Let the semantic differences between two entities be approximated by the length of the path that must be traversed to move between the points representing those entities. If one designs a COBOL-directed architecture A1, the semantic gap between this architecture and the COBOL language is reduced, but the gap between this architecture and the other languages may be found to be greater than the gap between these languages and the conventional von Neumann architecture.

There are two apparent solutions to this dilemma. The first is to define the architecture to be to the right of the von Neumann architecture in Figure 11.1, but not so much that it becomes highly oriented to a particular language. In other words, the architecture would incorporate the semantic aspects that the languages have in common. The second solution is to define the system such that it has multiple architectures (e.g., via multiple processors or by dynamically reconfiguring its architecture), where each architecture is directed toward a particular language.

The Burroughs B1700 computer series is an example of the second approach.

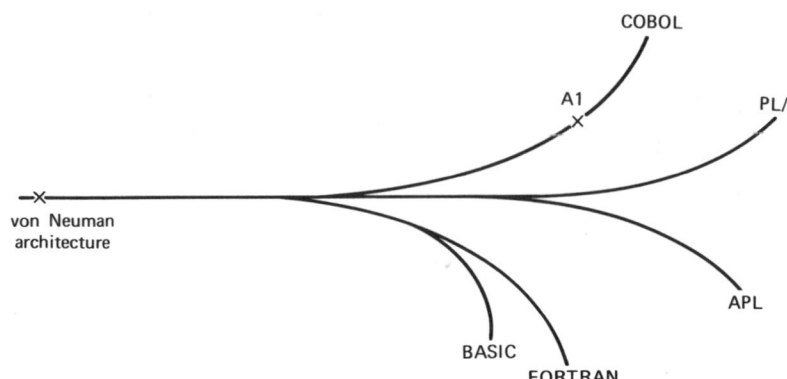

Figure 11.1 Another view of the semantic gap.

B1700 SYSTEM ARCHITECTURE

Unlike the systems discussed in previous chapters, the B1700 is a real commercial system. It is a small- to medium-scale system, supporting the BASIC, FORTRAN, COBOL, and RPG II programming languages. The B1700 is a compatible family of six processors, differing in speed and memory capacity. A more recent family, the B1800 series, has the same architecture, but it employs more advanced memory and logic technologies. The B1700 system is a multiprogramming, virtual-storage system and supports data base and communications applications.

The best way to understand the B1700 is to analyze its multiprogramming operations. Assume that a COBOL program and a FORTRAN program are currently being processed and that the system's operating system (the MCP—Master Control Program) is currently executing and is in the process of determining which application program is to be resumed.

The MCP is written in SDL (System or Software Development Language), a language similar to ALGOL and PL/I. In deciding on the optimal architecture to execute the MCP, the architects invented an architecture that is highly oriented to SDL. Hence, at this instant, the processor has an SDL-oriented architecture.

The architect's recognized, however, that this SDL-oriented architecture would be a poor execution target for application programs (i.e., a large semantic gap would exist between this architecture and BASIC, FORTRAN, COBOL, and RPG programs). To solve this problem, they defined multiple architectures, each oriented toward a specific language. Each architecture is embodied in a distinct microprogram. If the MCP decides that execution of the COBOL program should be resumed for some quantum of time, it directs the underlying machine to switch microprograms, in this case to the microprogram giving the machine a COBOL-oriented architecture. This act completely changes the system's computer architecture (e.g., giving the processor a totally different instruction set and data formats).

The machine now resumes execution of the COBOL program, compiled to this COBOL-oriented architecture. (Each of these distinct architectures is called an *S-language*). Assume now that an I/O interrupt occurs. The current microprogram switches to the SDL-oriented microprogram, and the MCP begins executing. If the MCP decides to resume execution of the FORTRAN program, it recognizes that it has been compiled to a different S-language; it instructs the current microprogram to switch to the FORTRAN-oriented microprogram, giving the processor a FORTRAN-oriented architecture.

Thus the B1700 dynamically changes its architecture as the MCP dispatches different application programs. The B1700 has a distinct S-language for each programming language, with the exception of COBOL and RPG. Because of the semantic similarities between COBOL and RPG, programs in both languages are compiled to the same S-language.

IMPLEMENTATION CONSIDERATIONS

Although the subject of processor design is purposely excluded from this book, the "multiarchitecture" nature of the B1700 presents several problems in the implementation of the underlying processor. It is worthwhile to consider briefly how these problems are solved.

One problem is how to select the optimal memory-word size and data-path width, since these are normally closely related to the data sizes in the architecture, but the B1700 has multiple unrelated architectures. For instance, perhaps the optimal word size is 18 or 36 bits when the processor has a FORTRAN-oriented architecture; however, this might prove unwieldy when the processor has a COBOL-oriented architecture, where one is more likely to desire a decimal, byte-oriented machine.

The solution to this problem was to allow the apparent physical characteristics of the processor, as well as the architecture, to vary dynamically. For instance, the width of the processor's ALU (arithmetic and logical unit) can vary dynamically. When the processor has a FORTRAN-oriented architecture, the ALU might appear to be 18 bits wide, but when the processor has a COBOL-oriented architecture, the ALU might have the appearance of being 8 (or some multiple thereof) bits wide. This is accomplished by an internal register in which the microprogram can store the desired width of the ALU. If a microprogram stores the value 16 in this register, the ALU performs operations on 16-bit operands. The ALU has a fixed physical width (24 bits), but this register masks off the unwanted bits in the ALU, creating the illusion of an ALU whose width can vary dynamically.

The same problem occurs with memory; that is, at one time one may want to address memory as if it had a word size of 18 bits, but at another time as if it had a word size of 8 bits. The solution here was to make memory bit addressable and to include a field-length specification in the memory-processor interface. For instance, if a microprogram wants to address memory in 16-bit units, when the microprogram sends a write request to memory it specifies a bit address, sets the field-length to 16 bits, and transmits a 16-bit value.

Another consideration is the apparent need for a large amount of control storage to hold the multiple microprograms. First, the problem is not as significant as it first appears to be, for the primary portion of each microprogram (excluding debugging aids) occupies approximately 28,000 bits of storage (less than 2000 16-bit microinstructions). Second, the system designer is presented with several alternatives. A microinstruction is provided that can move an area from main storage into an area in control storage, allowing the designer to store the microprograms in main storage and page them into control storage as needed. Also, the designer can direct the processor to fetch microinstructions directly from main storage, although the machine operates more slowly in this mode because of the longer access time of main storage. (The B1800 has no control storage; instead, microprogram segments are demand paged into a high-speed cache memory. Also, the low-end processors of the B1700 and B1800 families have no control storages or cache memories; microinstructions are fetched from main memory.)

STORAGE AND PERFORMANCE

As mentioned in Part I, a significant result of a higher-level architecture is the significant reduction in the number of bits needed to represent a program. This result is dramatically displayed on the B1700. As one example [1], seven sample FORTRAN programs occupied 280K bytes on the B1700, 560K bytes on the IBM System /360, and 450K bytes on the Burroughs B3500, an improvement of 50% and 38%, respectively, in memory utilization. Twenty sample COBOL programs were found to occupy 450K bytes on the B1700 and 1490K bytes on the System /360, an improvement of 70%. A sample of 31 RPG II programs occupied 150K bytes on the B1700 and 310K bytes on the IBM System /3, an improvement of 52%.

Part I also attributed significant performance advantages to the idea of higher-level architectures. As evidence, a compute-bound RPG program was found to run in 25 seconds on the B1700 and 208 seconds on an IBM System /3 model 10, a factor of 8:1 [1]. However, the B1700 had a cost of approximately 75% more, yielding a cost-performance differential of about 5:1. Although the average instruction time of the B1700's RPG-oriented architecture is about 35 microseconds, versus 6 microseconds for the System /3, significantly fewer instructions (about 50 : 1) were executed on the B1700.

REFERENCES

1. W. T. Wilner, "Design of the Burroughs B1700," *Proceedings of the 1972 Fall Joint Computer Conference.* Montvale, N.J.: AFIPS, 1972, pp. 489–497.
2. W. T. Wilner, "Burroughs B1700 Memory Utilization," *Proceedings of the 1972 Fall Joint Computer Conference.* Montvale, N.J.: AFIPS, 1972, pp. 579–586.
3. W. T. Wilner, "Microprogramming Environment on the Burroughs B1700," *Digest of the 1972 IEEE CompCon.* New York: IEEE, 1972, pp. 103–106.
4. *Burroughs B1700 System Reference Manual.* 1057155, Burroughs Corp., Detroit, 1972.
5. *Burroughs B1700 Systems for Computer Science Education.* 1077435, Burroughs Corp., Detroit, 1974.

12 | Burroughs B1700 COBOL /RPG Architecture

Because the B1700 dynamically changes its computer architecture, a discussion of the computer architecture of the B1700 is impossible. Rather than discussing all its architectures, we examine one: the architecture to which COBOL and RPG programs are compiled [1].

The architecture was designed to minimize the semantic gap between it and the COBOL and RPG languages; as a result, most COBOL statements and operators map into single machine instructions. In terms of the concepts discussed in Chapter 4, the architecture incorporates self-identifying (tagged) data, descriptors of composite data objects, pushdown stacks for subroutine management, variable-size storage cells, and decimal arithmetic. Its instruction set is neither stack or register oriented; a form of storage-to-storage addressing is used to reference operands. Lexical-level addressing is not used (it would be of no value for COBOL and RPG programs); instead, storage is addressed via a storage-segment-name /displacement couple, a form of addressing that is common across most Burroughs' systems and is a primitive forerunner of capability-based addressing.

Unlike the architectures in the previous chapters, the B1700 COBOL /RPG architecture is not described via examples; the documentation available on the memory concepts of the architecture is not quite detailed enough to illustrate the representation of a COBOL or RPG program.

DATA TYPES

Four basic data types are provided: an unsigned string of 4-bit decimal digits, a signed string of 4-bit digits, an unsigned string of 8-bit characters, and a signed string of 8-bit characters. If the string is signed, the sign (+ or −) is represented in the first 4 or 8 bits. When unsigned data types are used in numerical operations, their values are assumed to be positive. When an unsigned data type is used as the target of a numerical operation, the absolute value of the result is stored.

Four-bit cells (strings of 4-bit digits) always have a numerical value. The value is represented in binary-coded decimal. The maximum size of a 4-bit cell is 4095 digits, excluding the sign.

Eight-bit cells (strings of 8-bit characters) use the EBCDIC representation (data can also be represented in ASCII, but discussion of this is excluded). If the cell has a numerical value (i.e., all characters except for the optional sign have the value 1111xxxx, where xxxx is a binary-coded-decimal value), arithmetic operations may use it as an operand. The maximum size of an 8-bit cell is 1023 characters, excluding the optional sign.

Since the data types are self-identifying, the machine instructions are generic, and automatic data conversions are performed. For instance, one source operand of an ADD instruction could be a signed 4-bit cell of four digits, the other operand an unsigned 8-bit cell of five characters (providing it has a numerical value), and the target operand a signed 8-bit cell of seven characters.

The last type is an array: a homogeneous collection of the four data types. An array can have one, two, or three dimensions. For arrays of 4-bit cells, the maximum size is 524,287 digits. For arrays of 8-bit cells, the maximum size is 262,143 characters. The machine provides automatic subscripting and indexing. When an instruction indicates that one of its operands is an array, the machine automatically locates the appropriate element. The instruction can specify the subscript values explicitly (the subscript values are operands of the instruction) or implicitly (the subscript values are stored with the array descriptor).

PROGRAM PARAMETERS

One unique feature of this architecture is that the sizes of fields stored within instructions and descriptors (e.g., storage addresses) can vary from program to program, although their sizes are fixed for any particular program. This solves a critical problem facing computer architects: determining the size of fields, such as address fields. If, in an architecture with fixed-size addresses, the address-field size is large, storage is wasted in programs that do not need the maximum addressing range. On the other hand, if address fields are short, the addressing range of programs is restricted, and complicated mechanisms are needed to overcome it (e.g., the use of multiple base registers in the IBM System/370).

In this architecture, the compiler selects the smallest field sizes that are large enough to represent the particular object program and then compiles the program using these field lengths and communicates these field lengths to the processor by storing parameters with the object program. These program parameters are listed in Table 12.1. The purpose of the parameters is discussed in later sections. Table 12.1 shows the size in bits of each parameter and the range of valid values for the parameters. All parameter values are unsigned positive binary numbers.

To illustrate two of the parameters, memory addresses consist of a segment name and a displacement. If, in a particular program, PSEG=00110 (6) and PDISP=01100 (12), addresses in that program have 6-bit segment names and 12-bit displacements. (This type of address, however, does not appear in machine instructions.)

STORAGE STRUCTURE

Each object program is represented by a data space, as shown in Figure 12.1. The data space is located via a base register and contains the program's status information and data areas. The machine instructions

Table 12.1 Program Parameters

Parameter	Abbreviation	Size	Value Range
Data-segment-name size	PSEG	5	0–18
Data-displacement size	PDISP	5	1–21
Cell-length size	PCELL	5	1–14
Descriptor-index size	PINDEX	5	1–31
Deseriptor-entry size	PDESC	6	6–57
Branch-displacement size +1	PBDISP	5	2–31
Subroutine-stack size	PSSIZE	5	1–31
Offset of data segment zero	PDSEGZ	24	Any
Offset of subroutine stack	PSTACK	24	Any
Offset of descriptor table	PDTAB	24	Any

STORAGE STRUCTURE

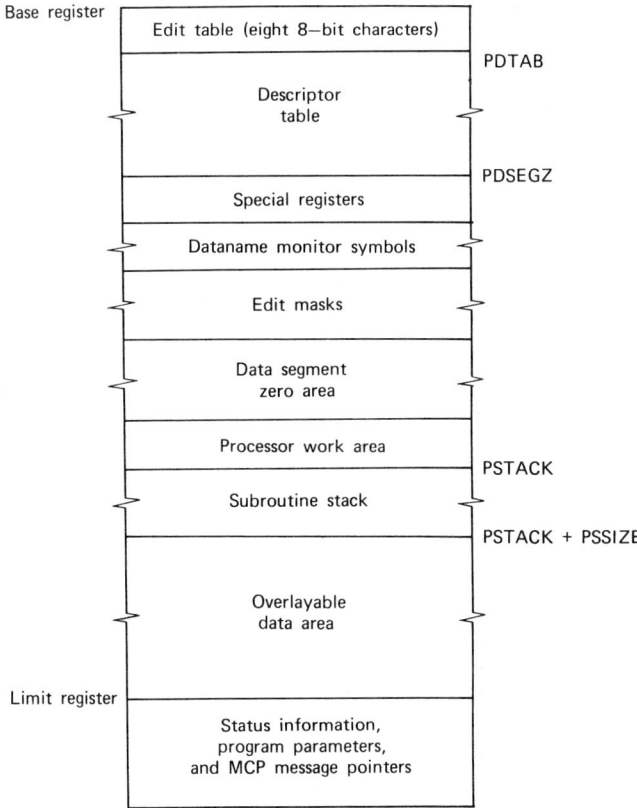

Figure 12.1 B1700 COBOL data space.

and space for the program's variables are in two or more separate memory segments.

The edit table contains eight special characters used by some of the editing instructions. The descriptor table contains a data descriptor for each variable in the program (it is analogous to the name table in SYMBOL). The edit-masks area contains strings of editing subinstructions that are referred to by one of the editing instructions. The data-segment-zero area holds program-control information and data fields that are used when communicating with the MCP. The subroutine stack holds state information about each active subroutine.

The descriptor table (called the COP or current operand table by Burroughs) contains an entry for each program variable. Machine instructions address their operands by specifying indexes of descriptors in this table; the descriptors in turn point to the actual data locations. The index of the first table entry is 1.

The format of a descriptor for a nonarray cell is shown in Figure 12.2. The first two fields contain the memory address of the associated data cell: the name of the segment containing the cell and the displacement (in units of 4 bits) of the first bit of the cell within the segment. The third field defines the length of the cell (excluding its sign, if one is present) in 4- or 8-bit units, depending on the type of cell. The displacement and length fields contain unsigned binary values. In this descriptor the SIF bit (subscript-index flag) is off, indicating a nonarray cell.

The data-type field specifies the cell's type:

00 = unsigned 4-bit
01 = unsigned 8-bit
10 = signed 4-bit
11 = signed 8-bit

The ASCII flag specifies whether the cell's value is stored in the ASCII or EBCDIC representation. This flag influences the execution of certain machine instructions, but hereafter we discuss only EBCDIC data.

Array cells are represented by an extended descriptor, as shown in Figure 12.3. Array descriptors are stored in multiple descriptor-table entries. The seventh field specifies the number of dimensions (subscripts or indexes). The SI field specifies whether subscripting or indexing is applicable. (For a definition of the distinction between subscripting and indexing, refer to a COBOL manual.)

If subscripting is indicated, one to three fields (depending on the number of dimensions) are present and contain the binary value by which each subscript value is to be multiplied to locate the appropriate element. The table bound specifies the displacement of the last element in the array. If the computed element location is greater than the table bound, an error is generated. (Hence for a 3 × 4 array, the machine

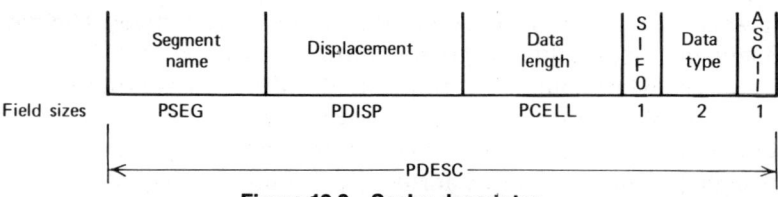

Figure 12.2 Scalar descriptor.

INSTRUCTION FORMATS

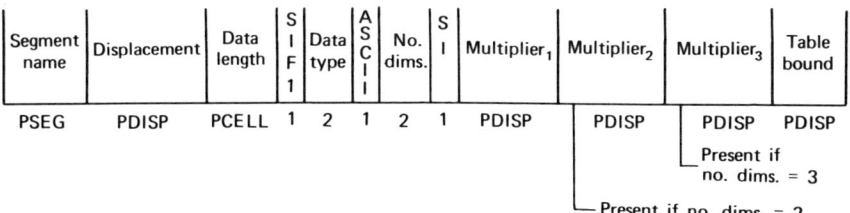

Figure 12.3 Array descriptor.

would not detect a reference to element 4,2 as an error, but it would detect a reference to element 4,4).

For a subscripted or indexed array, descriptors for each subscript or index value (one to three) must immediately follow the array descriptor. These descriptors must be scalar descriptors. Hence, given a reference to an array descriptor, the processor can locate the current subscript or index values to compute the address of the current array element.

INSTRUCTION FORMATS

Each machine instruction consists of an operation code followed by a variable number of fields, the number depending on the instruction type. These fields are usually literal values or descriptor indexes. Opcodes are encoded into 3 or 9 bits. The seven most frequent instructions (INCREMENT, MOVE ALPHANUMERIC, MOVE NUMERIC, BRANCH, PERFORM ENTER, COMPARE ALPHANUMERIC, and COMPARE NUMERIC) have 3-bit op-codes; the remaining instructions have 9-bit op-codes with the format 111xxxxxx.

With a few exceptions that are discussed later, the remaining fields in an instruction can be a descriptor reference (DR), a literal or descriptor reference (LIT/DR), or a branch address (BA). These formats are shown in Figure 12.4. For instance, a DR can be either an index to an entry in the descriptor table or an inline descriptor. Descriptor indexes are unsigned binary values. Inline descriptors for nonarrays have the same format as descriptors in the descriptor table, but inline descriptors for arrays have a slightly different format. Rather than repeating the descriptors for the subscript values after the inline array descriptor, the inline array descriptor contains descriptor indexes to the subscript descriptors.

A LIT/DR field can contain a literal value or a descriptor reference. For literals of less than eight digits or characters (excluding the sign),

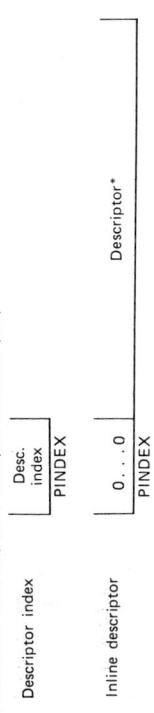
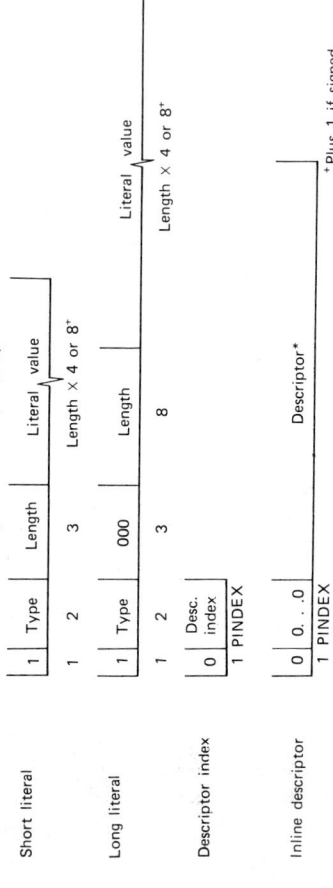
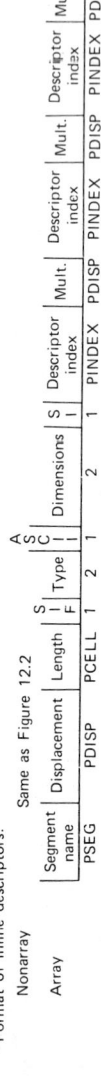
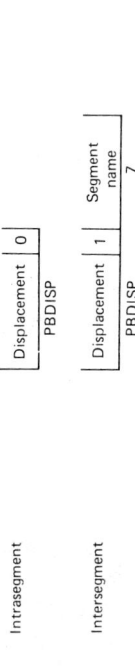

Figure 12.4 Instruction field formats.

MACHINE INSTRUCTIONS

the short form is used. For longer literals, the long form is used. A literal can be a signed or unsigned 4-bit data type or an unsigned 8-bit data type.

A BA (branch address) is used in branching instructions. If the branch is to a location within the current code segment, the intrasegment form is used. The displacement value is an unsigned binary value specifying the bit displacement of an instruction relative to the beginning of the segment.

MACHINE INSTRUCTIONS

The architecture contains 38 instructions that can be logically grouped into four categories: arithmetic, data movement, control, and miscellaneous. The seven arithmetic instructions accept any of the four data types as operands; automatic data conversions and padding with leading zeros are performed where needed. If an 8-bit data type is used as a source operand of an arithmetic instruction, it must have an EBCDIC numeric value (e.g., the values F0F5 (5) and 43F3F2 (+32) are valid, but the value A1BC is not).

The 11 data-movement instructions move the value of one cell into another, while performing implicit or explicit editing operations. The 16 control instructions are associated with altering the sequential flow of execution of machine instructions. The four miscellaneous instructions are intended for communication with the MCP (e.g., for I/O operations).

Arithmetic Instructions

Instruction: INCREMENT (Two-address add)
Function: The first operand is added to the second operand.
Format: OP,LIT/DR,DR

Instruction: ADD (Three-address add)
Function: Two operands are added, and the result is placed in the third operand.
Format: OP,LIT/DR,DR,DR
Operands: Operands 1 and 2 may specify the same cell, but operand 3 must be a different cell than operands 1 and 2.

Instruction: INCREMENT-BY-ONE
Function: The operand is incremented by 1.
Format: OP,DR

Instruction: SUBTRACT

Function: The first operand is subtracted from the second operand, and the result is stored in the third operand.

Format: OP,LIT/DR,DR,DR

Instruction: DECREMENT

Function: The operand is decremented by 1.

Format: OP,DR

Instruction: MULTIPLY

Function: Two operands are multiplied, and the product is stored in the third operand.

Format: OP,LIT/DR,DR,DR

Operands: The third operand must be a signed or unsigned 4-bit data type.

Instruction: DIVIDE

Function: The second operand is divided by the first operand. The quotient is stored in the third operand; the remainder is stored in the second operand.

Format: OP,LIT/DR,DR,DR

Operands: The sign of the remainder is that of the dividend (the second operand). A division by zero sets the overflow condition.

Data-Movement Instructions

Instruction: MOVE ALPHANUMERIC

Function: The first operand is moved into the second operand. If the first operand is a 4-bit data type, the transferred value is converted to EBCDIC.

Format: OP,LIT/DR,DR

Operands: If the length of the second operand (target) is greater than that of the first (source), the target is padded on the right with spaces (if it is an 8-bit data type) or zeros (if it is a 4-bit data type). If the target length is less than the source length, truncation on the right occurs. If the source is a 4-bit data type, each digit is stored in the target in the form 1111xxxx (with the exception of the sign, if present, which is converted to its proper EBCDIC value).

MACHINE INSTRUCTIONS

Instruction: MOVE NUMERIC

Function: The first operand is moved into the second operand.

Format: OP,LIT/DR,DR

Operands: Data conversion is performed if necessary. If the target operand is unsigned, the sign of the source is ignored. If the target is larger than the source, the target is padded on the left with zeros. If the target is smaller than the source, the data is truncated from the left.

Instruction: MOVE SPACES

Function: The operand is filled with EBCDIC spaces (01000000).

Format: OP,DR

Operands: The operand must be an unsigned 8-bit data type.

Instruction: MOVE ZEROS

Function: The operand is filled with zeros.

Format: OP,DR

Operands: If the operand is signed, the sign is set to positive.

Instruction: MOVE TRANSLATE

Function: The first operand is moved into the second operand, translating each character as it is moved via a specified translate table.

Format: OP,LIT/DR,DR,DR

Operands: The first and second operands must be unsigned 8-bit data types. The third operand is used as a translation table. Translation is accomplished by using each source character, multiplied by 8, as an index into the table. The table character at the computed address is moved to the target. If the target is larger than the source, the target is padded on the right with spaces. If the target is smaller than the source, the data is truncated on the right.

Instruction: SCALED MOVE NUMERIC

Function: A move-numeric operation is performed, moving the first operand into the second operand. A scale factor is first added to or subtracted from the length of the source operand.

Format: OP,LIT/DR,DR,V,SCL

Operands: The V field is a single bit. The SCL field is an unsigned binary value of PCELL bits in length. If V=0, SCL (the scale factor) is added to the length of the source operand, and the extended portion of the source field is assumed to be zeros on the right. SCL must not be greater than the target length. If V=1, SCL is subtracted from the length of the source operand. SCL must not be greater than the source length. After adjusting by the scale factor, the rules of the MOVE NUMERIC instruction apply.

Instruction: CONCATENATE

Function: Each of the source operands is moved into the target operand.

Format: OP,N,DR,LIT/DR1,...,LIT/DRN

Operands: N is a 4-bit unsigned binary number specifying the number of source operands. The next field is the descriptor reference of the target operand. The remaining N fields specify the source operands. Data is moved by the rules specified for the MOVE ALPHANUMERIC instruction.

Instruction: EDIT

Function: The first operand is moved into the second operand. The move is controlled by a string of subinstructions at a specified location.

Format: OP,LIT/DR,DR,DADDR

Operands: DADDR is a PDISP-bit unsigned binary value that specifies the digit displacement of the string of subinstructions relative to the PDSEGZ (data segment zero) base (the "edit masks" area in Figure 12.2). If the target is larger than the source, the target is padded on the left with zeros. If the target is smaller than the source, the data is truncated from the left. The source operand can be any data type, but the target must be an unsigned 8-bit data type.

Subinstructions: The edit subinstructions are defined in Table 12.2. The value nnnn is a 4-bit binary value used as a repeat count. For instance, the value 0000 means no repeat (one time only); the value 0010 means repeat execution of the subinstruction twice. The value xxxx is a 4-bit unsigned binary value used to skip over a number of target characters. The value 0000 means no skip. The value yyyy is used to index a table of editing constants (Table 12.3). If yyyy=1010, the editing character follows the yyyy in storage, and the next subinstruction follows the character.

During an edit operation, the processor maintains three switches: S (sign), Z (zero suppress), and P (check protect). At the start of the EDIT instruction, Z and P are set to zero, and S is set to zero if the source operand is positive (or unsigned), otherwise to 1.

MACHINE INSTRUCTIONS

Table 12.2 EDIT Subinstructions

Subinstruction	Function
0000 nnnn	Move digits
0001 nnnn	Move characters
0010 nnnn	Move suppress
0011 nnnn	Fill suppress
0100 xxxx	Skip reverse destination
0101 yyyy	Insert unconditionally
0110 yyyy	Insert on minus
0111 yyyy	Insert suppress
1000 yyyy	Insert float
1001 yyyy	End float mode
1010 0000	End nonzero
1010 0001	End of mask
1010 0010	Start zero suppress
1010 0011	Complement check protect
others	Undefined

Table 12.3 EDIT Constants

Table Entry	Inserted Character
0	+
1	−
2	*
3	.
4	,
5	$
6	0
7	Blank
8	Either + or −
9	Either blank or −
10	Inline 8-bit character

Subinstruction: MOVE DIGIT

Operation: Set Z to 1. Move the next source digit or character to the next location in the target. If a digit (4-bit unit) is being moved, convert it to an EBCDIC character. If a character is being moved, set its first 4 bits to 1111.

Subinstruction: MOVE CHARACTER

Operation: Set Z to 1. Move the next source digit or character to the next location in the target. If a digit is being moved, convert it to an EBCDIC character.

Subinstruction: MOVE SUPPRESS

Operation: If the next source digit or character does not have the value zero, or if Z=1, a move-digit operation is performed. If not, (1) if P=0, a blank is stored in the next target location or (2) if P=1, an asterisk (*) is stored.

Subinstruction: FILL SUPPRESS

Operation: If P=0, store a space (blank) in the next target location. Otherwise, store an asterisk. The source-location counter is not incremented.

Subinstruction: SKIP REVERSE DESTINATION

Operation: Subtract nnnn from the next-location counter to the target operand. The source-location counter is not incremented.

Subinstruction: INSERT UNCONDITIONALLY

Operation: Store the designated edit-table or inline character in the next target location. If table-entry 8 is indicated, store a + if S=0 or a − if S=1. If table entry 9 is indicated, store a blank if S=0 or a − if S=1. The source-location counter is not incremented.

Subinstruction: INSERT ON MINUS

Operation: Store the designated character in the next target location, based on the following conditions. The source-location counter is not incremented.

S=1 T=0 . . . 7	store table-entry T
S=0 P=0	store blank
S=0 P=1	store *
S=1 T=8	store −
S=1 T=9	store −
S=1 T=10	store inline character

Subinstruction: INSERT SUPPRESS

Operation: Store the designated character in the next target location, based on the following conditions. The source-location counter is not incremented.

Z=1 T=0 . . . 7	store table-entry T
Z=0 P=0	store blank
Z=0 P=1	store *
Z=1 S=0 T=8	store +

MACHINE INSTRUCTIONS

Z=1 S=1 T=8	store −
Z=1 S=0 T=9	store blank
Z=1 S=1 T=9	store −
Z=1 T=10	store inline character

Subinstruction: INSERT FLOAT

Operation: Store the designated character and/or perform a move-digit operation on the next source digit or character in the next, or next two target locations, based on the following conditions. (SRC = next source digit or character).

Z=1	move digit
Z=0 SRC=0 P=0	store blank
Z=0 SRC=0 P=1	store *
Z=0 SRC≠0 T=0 . . . 7	store table entry T, then move digit
Z=0 SRC≠0 T=8 S=0	store +, then move digit
Z=0 SRC≠0 T=8 S=1	store −, then move digit
Z=0 SRC≠0 T=9 S=0	store blank, then move digit
Z=0 SRC≠0 T=9 S=1	store −, then move digit
Z=0 SRC≠0 T=10	store inline character, then move digit

Subinstruction: END FLOAT MODE

Operation: Store the designated character in the next target location, based on the following conditions. The source-location counter is not incremented.

Z=0 T=0 . . . 7	store table entry T
Z=0 T=8 S=0	store +
Z=0 T=8 S=1	store −
Z=0 T=9 S=0	store blank
Z=0 T=9 S=1	store −
Z=0 T=10	store inline character
Z=1	no operation

Subinstruction: END NONZERO

Operation: Terminate the EDIT instruction if any nonzero source digit or character has been moved. Otherwise, continue with the next subinstruction.

Subinstruction: END OF MASK

Operation: Terminate the EDIT instruction.

Subinstruction: START ZERO SUPPRESS
Operation: Set Z=0.

Subinstruction: COMPLEMENT CHECK PROTECT
Operation: Complement the value of P.

Instruction: EDIT WITH EXPLICIT MASK
Function: The first operand is moved into the second operand. The move is controlled by a string of subinstructions within the instruction.
Format: OP,LIT/DR,DR, subinstruction string
Operands: The format of subinstruction string is the same as for a literal (Figure 12.4). The first 2 bits (type) must be set to 01. If the next 3 bits (length) are not zero, they indicate the length of the literal that follows (a string of subinstructions). If the 3 bits are zero, the next 8 bits designate the length of the literal that follows (a string of subinstructions). The rules of the EDIT instruction apply.
Subinstructions: Same as those of the EDIT instruction. The difference between the two instructions is that the subinstructions are included inline with the EDIT-WITH-EXPLICIT-MASK instruction, but are stored separately when the EDIT instruction is used.

Instruction: MICR EDIT
Function: The first operand is moved into the second operand. Certain characters are deleted (not moved), and a count of all characters moved is placed in the third operand.
Format: OP,DR,DR,DR
Operands: This instruction is intended for use with magnetic-ink-character readers; the details of its operation are not discussed here.

Instruction: MICR FORMAT
Function: The first operand is moved and formatted into the second operand.
Format: OP,DR,DR
Operands: This instruction is intended for use with magnetic-ink-character readers; the details of its operation are not discussed here.

Control Instructions

Instruction: BRANCH
Function: Control is transferred to the instruction at a specified address.
Format: OP,BA (see Figure 12.4)

MACHINE INSTRUCTIONS

Instruction: COMPARE ALPHANUMERIC

Function: The first operand is compared to the second operand in a binary fashion. If the specified relation is true, control is transferred to the instruction at a specified address.

Format: OP,LIT/DR,DR,R,BA

Operands: R is a 3-bit field indicating the following relations: 001 (>), 010 (<), 011 (≠), 100 (=), 101 (>or =), and 110 (< or =).

Instruction: COMPARE NUMERIC

Function: The first operand is compared to the second operand, based on their algebraic values. If the specified relation is true, control is transferred to the instruction at a specified address.

Format: OP,LIT/DR,DR,R,BA

Operands: R is defined as in the COMPARE-ALPHANUMERIC instruction.

Instruction: COMPARE REPEAT

Function: The first operand is repeatedly compared in a binary fashion to successive subsets of the second operand. If the specified relation is true, control is transferred to the instruction at a specified address.

Format: OP,LIT/DR,DR,R,BA

Operands: The two operands must be unsigned 8-bit data types. The length of the second operand must be an integral multiple of the length of the first operand. R is defined as in the COMPARE-ALPHANUMERIC instruction.

Instruction: COMPARE FOR SPACES

Function: The operand is compared in a binary fashion to a field of all spaces. If the specified relation is true, control is transferred to the instruction at a specified address.

Format: OP,DR,R,BA

Operands: The operand must be an unsigned 8-bit data type. R is defined as in the COMPARE-ALPHANUMERIC instruction.

Instruction: COMPARE FOR ZEROS

Function: The operand is algebraically compared to a field of all zeros. If the specified relation is true, control is transferred to the instruction at a specified address.

Format: OP,DR,R,BA

Operands: R is defined as in the COMPARE-ALPHANUMERIC instruction.

Instruction: COMPARE FOR CLASS

Function: The format of the operand is analyzed. If it matches the specified format, control is transferred to the instruction at a specified address.

Format: OP,DR,C,BA

Operands: C is a 2-bit field indicating the desired test: 00 (completely alpha), 01 (completely numeric), 10 (not completely alpha), 11 (not completely numeric). The operand must be an 8-bit data type.

Instruction: BRANCH ON OVERFLOW

Function: The overflow condition is tested, and, if it matches the specified condition, control is transferred to the instruction at a specified address.

Format: OP,V,BA

Operands: V is a 1-bit field. If V=1 and the overflow condition is present, or if V=0 and the overflow condition is not present, the branch is taken.

Instruction: SET OVERFLOW

Function: The overflow condition is explicitly set or reset.

Format: OP,V

Operands: V is a single bit. Its value is stored in the overflow condition. (The overflow condition is implicitly set as a result of a division by zero.)

Instruction: GO TO DEPENDING ON

Function: Control is transferred to one of several locations, depending on the value of the operand.

Format: OP,DR,N,BA0,BA1, . . ., BAN

Operands: N is a 10-bit unsigned binary value. If the operand is less than one or greater than N, the branch address is BA0. If the operand is in the range one to N, the operand is used as an index to obtain the corresponding branch address.

Instruction: ALTER

Function: An address constant is copied into a specified data area.

Format: OP,DADDR,BA

Operands: DADDR is a PDISP-bit unsigned binary value that specifies

MACHINE INSTRUCTIONS

the digit displacement of an area relative to the PDSEGZ (data segment zero) base. BA (a branch address) is stored in the specified area.

Instruction: ALTERED GO TO PARAGRAPH

Function: Control is transferred to the instruction at a specified address. The address is obtained from an address constant in storage.

Format: OP,DADDR

Operands: DADDR is a PDISP-bit unsigned binary value that specifies the digit displacement of an area relative to the PDSEGZ (data segment zero) base. The address constant (BA) at this location specifies the location to which control is transferred.

Instruction: PERFORM ENTER

Function: An entry containing the current instruction-segment name, the displacement of the next instruction, and a constant is pushed onto the subroutine stack, and control is transferred to the instruction at a specified address.

Format: OP,K,BA

Operands: K, the constant, is an 8-bit value.

Instruction: PERFORM EXIT

Function: Control is returned to the instruction designated by the top subroutine-stack entry, providing that the constant value in the entry matches a specified value. If not, control proceeds to the instruction following the PERFORM-EXIT instruction.

Format: OP,K

Operands: K, the constant, is an 8-bit value. If a match occurs, the top entry is removed from the stack.

Instruction: ENTER

Function: An entry containing the current instruction-segment name and the displacement of the next instruction is pushed onto the subroutine stack, and control is transferred to the instruction at a specified address.

Format: OP,BA

Instruction: EXIT

Function: Control is returned to the instruction designated by the top subroutine-stack entry. The top entry is removed from the stack.

Format: OP

Miscellaneous Instructions

Instruction: CONVERT

Function: The operand is converted to an unsigned 24-bit binary value and is stored in a specified location in data segment zero.

Format: OP,DR,DADDR

Operands: The operand must have a positive decimal value. Its value is converted to base two and stored at the location specified by DADDR. DADDR is a PDISP-bit unsigned binary value that specifies the digit displacement of the target area relative to the PDSEGZ (data segment zero) base.

Instruction: COMMUNICATE

Function: The length and absolute address of the operand are stored in a message-address buffer shared between this program and the MCP. Control is transferred to the MCP.

Format: OP,DR

Operands The length field of the operand is converted from a digit or character length to a bit length before being stored. The absolute bit address of the operand is also stored in the buffer.

Instruction: LOAD COMMUNICATE

Function: The last 24 bits of information in the message-address buffer shared between this program and the MCP is moved to a specified location in data segment zero.

Format: OP,DADDR

Instruction: MAKE PRESENT

Function: The specified data segment is loaded. The base relative address of the data area is placed in a specified location in data segment zero.

Format: OP,SEG,DADDR

Operands: SEG is a PSEG-bit field specifying a data-segment name.

REFERENCE

1. The material in this chapter is used by permission granted by the Burroughs Corporation. Most of the material is paraphrased from *B1700 COBOL/RPG–S-Language*, 1058823–015, Copyright 1973, Burroughs Corporation, Detroit, Michigan.

EXERCISES

12.1 If a COBOL program requires 43 entries in its descriptor table, what value for the PINDEX parameter (size of descriptor indexes) should be used by the compiler?

12.2 If a two-address add (INCREMENT) instruction is generated for this program, what would be the instruction's size, assuming the source operand is not a literal?

12.3 What would be the size of an INCREMENT instruction to add the value 2 to a variable?

12.4 Determine the meaning of the following MOVE-ALPHANUMERIC instruction. (The first 3 bits are the op-code; assume PINDEX=6).
0001000100111001101101 1

12.5 If the target of the above instruction is a signed 8-bit cell of length 3, what would be stored there?

12.6 Given the string of edit subinstructions 10100011 01010101 00100101 01010011 00000001 10100001 what would be the value of the 10-character target operand, given a source operand with hexadecimal value 00001045?

12.7 What are the most striking attributes of this COBOL-oriented instruction set? Why?

12.8 Take a small COBOL program and, on paper, compile it to this architecture (i.e., produce the descriptors, program parameters, and instruction stream). If you cannot because you do not know COBOL, learn it. How can one develop a computer architecture without understanding the most widely used programming language?

V
A Software-Reliability-Directed Architecture

13 | The SWARD Machine

The next case-study architecture, the SWARD (software-reliability-directed) machine [1], is an experimental architecture with a unique design motivation: to explore whether one can develop an unconventional architecture to substantially improve the reliability of programs (both application and system programs) executing on the machine. This motivation stems from the premise that the most significant problem in the computing field today is the inability to produce reliable programs. Much effort in the software-engineering field has been focused on this problem, such as quests for better software design, programming, and testing methodologies and tools, less error-prone programming languages, and mathematical proofs of program correctness. However, this effort has not completely solved the software-unreliability problem, and this is compounded by the increasingly critical nature of new system applications (e.g., aerospace programs, air-traffic-control systems, medical systems, online banking systems, real-time process control, and many others). Hence the problem was attacked from still another direction by determining how the underlying machine architecture might be designed to contribute to higher program reliability.

The idea of developing an architecture with the thought of enhancing program reliability is not totally new, but this architecture is the first known full-scale attack on the problem; prior work is much more limited in scope. One proposal [2, 3] was an experimental machine and operating-system architecture with the goal of detecting certain program errors (certain addressing errors). In this architecture, all accesses to storage are done indirectly through a descriptor that defines the storage area, its contents, and the access privileges (a form of capability-based addressing). However, for programs written in high-level languages, protection mechanisms such as this are useful only for a tiny fraction of all possible errors.

Ehrman [4], in an essay on architecture and debugging, illustrates the advantages of having the underlying machine recognize undefined data values, employ tags to make storage words self-identifying, and check the number and types of arguments in subroutine calls. Feustel [5] also points out the error detection and debugging advantages of an architecture with tagged storage. Others have suggested that the machine be primed with all valid module-call sequences to allow the machine to detect errors that cause modules to be invoked in an incorrect sequence [6], and having the machine maintain a trace of the last n instructions or branches executed to assist the debugging process [7].

DEVELOPMENT OF THE DESIGN GOALS

We could begin with a discussion of the SWARD architecture, but the need for many attributes of the architecture would not be apparent without first having an understanding of the underlying design objectives. Hence the first step is examining the design goals that led to the development of this particular architecture.

The central theme of the architecture is that it should enhance the reliability of the programs executing on it, under the assumption that these programs are written in today's common programming languages (e.g., COBOL, PL/I, FORTRAN) or similar languages. The ways in which the architecture might contribute to software reliability were grouped into five categories:

1. Semantic errors that are common to many programming languages and that are best detected by the underlying machine.
2. Program logic errors that can be detected or prevented by the machine.
3. Ways in which the machine can limit the consequences of software errors.
4. Ways in which the machine can encourage (or at least not discourage) good software design and programming practices.
5. Ways in which the machine can assist the testing and debugging processes.

For category 1, 27 classes of semantic errors were identified. It was

DEVELOPMENT OF THE DESIGN GOALS 181

determined that, in general, none of these classes of errors could be detected at compilation time. That is, the implication is not that the compiler could not theoretically detect some of these situations some of the time, but that the compiler cannot detect all occurrences of each error type. Rather than listing all 27 classes of errors, a few representative ones are listed.

1. A program attempts to access a variable whose value is currently undefined or unset. The variable might be a simple variable, an array element, a field in a structure, a pointer variable, an element in a string, and so on.
2. A program attempts to reference an array element where one of the subscripts is outside of the bounds of the corresponding dimension.
3. Through the use of pointer (reference) variables, a program addresses storage that is not part of its data space.
4. A program uses a pointer variable, but the referenced storage has been freed (the "dangling reference" problem).
5. A program causes a variable's value to have a type or attribute other than that expected by the compiler. For example, a program reads a file record and references the fields in the record by using a structure, but the physical representation of the record is inconsistent with the definition of the structure.
6. The units of storage allocation are smaller than the units of storage addressability, causing inaccurate address computations. For instance, in PL/I on the IBM S/370, fixed-length bit strings do not necessarily begin on byte boundaries, but addresses only refer to byte boundaries. If a program computes the address of a bit string and later refers to the string through this address, the wrong storage might be referenced. Similar problems occur in argument transmission (e.g., passing the address of a bit-string argument).
7. The number of arguments is unequal to the number of corresponding parameters.
8. The attributes /type /size of an argument differ from those of the corresponding parameter.
9. Two modules contain inconsistent declarations of a global variable.

One possible source of concern is whether these errors represent a significant number of actual software errors. All available evidence [1] is affirmative. For instance, in a sample of 39 errors taken from software produced by the Boeing Aerospace Company [8], 13 fall into the 27 classes of semantic errors. An analysis of 3361 programming errors found during system testing and operational use of a large command and control system [9] indicates that at least 487 of the errors fall into the 27 classes.

The second category of ways in which software reliability might be enhanced is the prevention or detection of certain types of program logic errors. (A semantic error occurs when a program has no defined meaning; a logic error occurs when a program may be syntactically and semantically correct, but it does not behave according to its specification.) Most logic errors, of course, cannot be detected by the underlying system, but 10 classes of logic errors that could be detected or prevented were identified. A few of them are

1. The data referenced by a pointer variable does not have the attributes expected by the compiler (e.g., a PL/I pointer, on which a data structure is based, is assigned the address of some different data structure and then used to reference the first data structure).
2. The program alters a variable whose value is supposed to be constant.
3. Data values become imprecise due to the constraints of fixed-size storage cells and/or base-two representations.
4. A subroutine alters an input argument (an argument whose value is not intended to be altered).

The third category is the ability to limit the consequences of errors when they do occur. Five objectives were listed for this category; all are associated with addressing and protection. For instance, one objective was that an error in a program (application or system) should not result in an accidental modification of another program or its data. Another objective was that an error in a module in a program should not result in a modification of any data that are not intended to be used by that module. That is, the address space of a module should be limited to its local variables, its parameters, and any global data that it explicitly defines or declares.

The fourth category was support of good software design and programming practices. Seven objectives were identified in this area, including:

1. Efficient support of highly modular programs.
2. Discouragement of the use of global data and the related practice of data scoping (implicitly sharing variables among nested proce-

EVALUATION OF CURRENT ARCHITECTURES

dures). Note that this objective implies that the architecture should *not* have lexical-level addressing.
3. Discouragement of assembly-language programming.

The last category consists of ways in which the machine might support (e.g., reduce the execution overhead of) testing and debugging aids and tools. Eleven testing and debugging functions that conceivably might be assisted by the underlying machine were identified, including:

1. Tracing the execution flow of program statements.
2. Tracing the execution flow of subroutine calls.
3. Producing a frequency map of statement executions.
4. Producing a map of branch directions executed.
5. Trapping or intercepting detected errors and performing some corrective action.

EVALUATION OF CURRENT ARCHITECTURES

Another step taken during the development of this architecture was to evaluate existing systems to determine whether the problem had already been solved. Since the vast majority of existing machines have von Neumann architectures, it would be reasonable to evaluate samples from this population. However, this would not be a productive step, since a typical von Neumann machine would fail to react to almost all the error situations. Hence five architectures were selected, three non-von Neumann architectures and two von Neumann architectures, but the latter were given the extra advantage of any run-time checks performed by the compilers.

The analysis was performed by identifying how each system would react to the 37 classes of semantic and logic errors. PL/I is a language in which all 37 error situations are possible; thus the analysis was performed by assuming that each error situation occurred in a PL/I program, determining how that PL/I construct would be mapped (compiled) into the architecture, and then determining how the architecture would react to the error. The detection of a few of the logic-error types was dependent on minor changes to the PL/I language; the assumption was made that these changes were present.

The three non-von Neumann architectures studied were the Burroughs B6500/6700, the Student-PL machine of Part II, and the SYMBOL machine of Part III. The two other systems were the IBM PL/I optimizing compiler on the S/370 and the IBM FORTRAN H compiler on the S/370. The analysis is biased in that the S/370 (a von Neumann machine) is given the assistance of software checks generated by the compilers, whereas the other three architectures are not.

For each error situation, each architecture was analyzed and placed into one of these categories:

1. The system prevents the error from ever occurring.
2. The system always detects the error.
3. The situation cannot arise.
4. Prevention or detection occurs, but it is an option that must be selected by the programmer.
5. Prevention or detection occurs, but it is an option that must be selected by the programmer, and even when selected it does not prevent or detect the error in all situations.
6. The system prevents the error from occurring in most, but not all, situations.
7. The system detects the error in most, but not all, situations.
8. The error is not prevented or detected in many or all situations.
9. The particular language construct cannot be implemented in the system.

Table 13.1 summarizes the results. Categories 1, 2, and 3 were treated as favorable outcomes, and the remaining categories were treated as negative outcomes. Table 13.1 indicates that even the non-von Neumann architectures do not sufficiently solve the problem, and PL/I on the S/370 consistently deals successfully with only one error type (division by zero).

Category 4 (error check is an option) was not counted as favorable because programmers are usually discouraged from using these checks because of time and space overhead (e.g., when the PL/I-S/370 checks

Table 13.1 Prevention/detection of 37 Error Types

Architecture	Score
B6500	11
Student-PL Machine	20
SYMBOL	24
PL/I on S/370	1
FORTRAN on S/370	5

were enabled, the compiler produced warning messages stating that the checks involve substantial overhead in both storage space and execution time). If category 4 was included in the favorable group, however, the score for PL/I on the S/370 would rise only to 3.

DEVELOPMENT OF THE ARCHITECTURE

This section presents some of the principal attributes of the architecture and the rationale for their existence. Chapter 14 discusses the storage concepts of the architecture and illustrates the architecture through several compilation examples. Chapter 15 is a specification of the instruction set.

A major concept in the architecture is that the machine should severely restrict the address space available to an individual module (e.g., FORTRAN subroutine or function subprogram, PL/I external procedure or function, COBOL subprogram). That is, a module's address space should be reduced to only those data to which the module needs access: its parameters, locally defined variables, and constants (i.e., only those data named in the source-language version of the module). The implication is that the machine must manage storage at a high level, much higher than the von Neumann view of a single linear sequential memory.

A second concept is related to the first concept: traditional machine addresses should be discarded. There are three types of storage objects that must be uniquely addressable: a module, an activation record (the collection of data allocated for an activation or invocation of a module), and an explicitly allocated block of storage. When one of these objects is created (i.e., a module is defined to the machine, a module is activated (called), a program explicitly allocates some storage), the machine assigns it a unique name (called a *logical address* and stored in an area called a *pointer*). The machine prohibits programs from creating logical addresses on their own and from altering the value of a pointer. When one of these objects is freed, its unique logical address is "never" reused. (Since a finite amount of storage is used to hold a logical address, only a finite number of logical addresses are available; thus they must eventually be reused. However, the size of a logical address can be adjusted to provide some "safe" period of time before a value is reused. In the SWARD machine, logical addresses are reused approximately every 2.5 months, assuming that a logical address is created every 100 microseconds.)

The instructions within a module can only address data defined within the module or data within any storage that is dynamically created by the module. Since a module cannot, on its own, create or alter a pointer, the only other storage it can reference is storage whose pointer is passed to the module from another module. Not only does this concept facilitate the detection of addressing errors (e.g., the dangling-reference problem), but it also serves as a storage protection mechanism. This addressing concept is the idea of capability-based addressing, as discussed in Chapter 4.

The third necessary concept in the architecture is that all data must be self-identifying. This means that descriptive information will be stored with each item of data, describing such attributes as its size and type. This self-identification allows the machine to detect incompatible operands of an operation and allows it to enforce other rules (e.g., the rule prohibiting the creation and manipulation of pointers). Two rules concerning self-identification, or tags, must be enforced: (1) the tag always describes the programmer's intended properties of the data (e.g., the attributes in the DECLARE statement), and (2) the value and representation of the data always agree with the tag.

In most other non-von Neumann machines, the concepts of tags and descriptors are treated distinctly. However, the concepts have much in common. In the SWARD machine the two concepts have been generalized into one concept called a tag.

To close the semantic gap between language data types and machine data representations, most data types known to the machine are variable in size. Not only does this prevent certain types of semantic errors that arise when variable-size language data types are mapped into fixed-size machine data types, but it leads to more efficient use of storage, as noted in Chapter 2.

Another vital step in closing the semantic gap is to eliminate the concept of binary (base 2) arithmetic and to represent all numbers in decimal (base 10) form. Although certain descriptive information in the architecture is represented in base 2, all data corresponding to variables in programs are represented in base 10.

One consideration that is influenced by many of the previous points is the method used by instructions to address their operands. Many forms have been proposed, but a basic underlying consideration is whether the architecture should contain general-purpose registers or evaluation stacks (or both or neither). In studying various addressing mechanisms, no apparent relationships to software reliability were found. However, the architecture contains neither registers nor evaluation stacks (but it does use stacks for subroutine linkages). An instruction addresses an operand by specifying the relative location of that operand within the address space of the module. Registers and evalua-

DEVELOPMENT OF THE ARCHITECTURE

tion stacks were not used for the reasons discussed in Chapter 4. In addition, the addressing method used in the architecture results in small address fields in instructions, negating the main advantage of register and stack-oriented instruction sets.

Another necessary concept in the architecture is the ability to distinguish between defined and undefined data values. In addition to its valid values, each data item should be able to have an additional value called "undefined." Any attempt to use a data item's value will be detected by the machine if the value is undefined. All addressable storage that is not initialized to some value is automatically initialized to "undefined." In addition, instructions are necessary to test explicitly a data item for an undefined value and to mark explicitly a data item as undefined (e.g., for a language in which the value of a loop iteration variable is supposed to be undefined when the loop terminates).

The "undefined" mark in this architecture is associated with the value or contents of a data item rather than with its tag because

1. "Undefined" is a property of the value of a data item rather than an attribute of a data item. A data item's tag remains constant, but its value changes during execution. Since the undefined state also may change during execution, it should be associated with the value of a data item.
2. For collections of data in which each data element is homogeneous (e.g., arrays and strings), it is desirable to store the tag once (with the collection), rather than storing tags with each element. Yet it is possible for some elements of a collection to be defined and others to be undefined, implying again that the undefined state is a value rather than a tag.

Given that the machine is aware of the concepts of modules and activation records and given that the machine must check arguments and parameters for consistent number and attributes, a logical deduction is that the architecture should provide a call/return mechanism that is semantically close to, or equivalent to, the CALL/RETURN statements in programming languages. That is, the CALL mechanism should allocate an activation record for the called module and add it to the stack of current activation records, initialize variables in the activation record, initialize parameters, suspend execution of the current module, and begin execution of the called module. Since all data in the system are tagged, the call mechanism needs a "die" for variables in the activation record, describing how each data item should be tagged when the activation record is created.

Since the architecture is supposed to detect such errors as exceeding an array dimension bound and inconsistent definitions of records (e.g., PL/I structures) among modules, the architecture must be aware of these data types. Hence the architecture contains the "less primitive" data types of arrays, structures (ordered trees of heterogeneous nodes), and strings. Supporting such data types is more involved than it first appears. For instance, a language such as PL/I provides for arrays of structures, structures of arrays, arrays of strings, structures of structures, structures of arrays of structures, and so on. A hopefully elegant representation of these is described in Chapter 14.

Note that the self-identification property mentioned earlier applies to these data types. For instance, every array, structure, and string is tagged. The machine instructions are generic; for instance, there is only one ADD instruction, and its two operands can be any meaningful data types that pass certain consistency tests. For instance, an operand to an ADD can be a simple numeric variable, an array element, an entire array, or a numeric field in a structure.

Another consideration in the architecture is a mechanism to handle exceptional conditions. The mechanism uniformly applies to any type of "fault" or interrupt, be it a machine detection of an error, detection of some explicitly identified event (i.e., for ON-units), or a machine detection of some debugging action such as the execution of a particular instruction. Each module is capable of specifying a special "fault-handling" entry point and capable of describing what types of faults can be handled by that entry point (somewhat analogous to the software mechanism in the CAL operating system [10]). When a fault occurs, the machine searches through the activation-record stack looking for the first module that wants to handle that type of fault. When one is found, the machine "calls" the entry point (making it a subroutine of the module initiating the fault) and passes it arguments describing the fault. A fault-handling entry point has the ability to resume execution at the point beyond the fault, repeat execution of the instruction causing the fault, or to decide to buck the fault up to a higher module. It is assumed that the highest module (the first one invoked in executing a program) is part of the operating system or a debugging tool, and this module will specify that it can handle all types of faults.

In relation to the concepts in Chapter 4, the SWARD architecture incorporates the ideas of self-identifying data, descriptors of composite data objects (although this has been generalized as a subset of the self-identifying-data concept), automatic subroutine management, capability-based addressing, decimal data representation, and variable-size storage cells. However, explicit decisions were made to exclude the ideas of lexical-level addressing and expression-evaluation stacks.

REFERENCES

1. G. J. Myers, "The Design of Computer Architectures to Enhance Software Reliability," Ph.D. dissertation, Polytechnic Institute of New York, 1977.
2. M. J. Spier, "A Model Implementation for Protective Domains," *International Journal of Computer and Information Sciences*, **2**(3), 201–299 (1973).
3. M. J. Spier, "A Pragmatic Proposal for the Improvement of Program Modularity and Reliability," *International Journal of Computer and Information Sciences*, **4**(2), 133–149 (1975).
4. J. R. Ehrman, "System Design, Machine Architecture, and Debugging," *SIGPLAN Notices*, **7**(8), 8–23 (1972).
5. E. A. Feustel, "On the Advantages of Tagged Architecture," *IEEE Transactions on Computers*, **C–22**(7), 644–656 (1973).
6. J. R. Kane and S. S. Yau, "Concurrent Software Fault Detection," *IEEE Transactions on Software Engineering*, **SE–1**(1), 87–99 (1975).
7. H. J. Saal and L. J. Shustek, "On Measuring Computer Systems by Microprogramming," *Microprogramming and Systems Architecture: Infotech State of the Art Report 23*. Berkshire, England: Infotech, 1975, pp. 473–489.
8. J. P. Rankin, G. J. Engles, and S. G. Godoy, "Software Sneak Circuit Analysis," AFWL-TR-75–254, Boeing Aerospace Co., Houston, Tex., 1976.
9. G. R. Craig et al., "Software Reliability Study," RADC-TR-74–250, TRW Systems Group, Redondo Beach, Calif., 1974.
10. J. Gray et al., "The Control Structure of an Operating System," RC-3949, IBM Research Div., Yorktown Heights, N.Y., 1972.

14 | Program Compilation and Execution on SWARD

The SWARD architecture is illustrated by examining the compilation and execution of several high-level-language programs. The machine is first defined by describing its data types, an important concept called a *module*, the instruction formats and addressing mechanisms, fault handling, and a summary of the instructions. Chapter 15 contains a detailed description of each instruction; the reader may wish to refer to Chapter 15 while studying the examples in this chapter.

DATA TYPES

Before discussing the data types, a few basic storage concepts must be introduced. The basic unit of storage allocation is a *token*, a 4-bit quantity. The basic unit of storage addressing is a *cell*, a variable number of contiguous tokens. A cell corresponds to a variable or data item in a source program and has two major components: a *tag*, which describes the attributes of the cell, and a *content*, which describes the value of the cell.

The machine recognizes 14 data or cell types of which 10 are considered *primitive* data types, one is a *structure* data type, and three are *nested* data types. The basic difference among the three categories is that primitive cells have single values, structure cells describe collections of other cells, and nested cells have tags which in turn contain tags.

DATA TYPES

Primitive Cell Types

The primitive cell types are decimal integer, decimal fixed point, decimal floating point, boolean, character, token, boolean string, character string, token string, and pointer. The tag of each cell describes its type and size; the contents component describes its value.

A *decimal integer* (di) cell has the format shown in Figure 14.1. All field lengths are expressed in tokens (4-bit units). The first field (1 token) indicates that this is a decimal integer. The second field is a binary number from 1 to 15 that indicates the number of decimal digits in the cell. The third field is the sign. If its last (fourth) bit is zero, the integer is positive; if the bit is a 1, the integer is negative. The last field holds the absolute value of the integer in binary-encoded decimal digits (0000–1001). The length of the last field is defined by the second field. The arrow is used in the figures to indicate the boundary between the tag and content components. Hence the decimal-integer cell has a 2-token (8-bit) tag and a 2–16-token content.

If the first value digit is 1111, the cell has the value "undefined." As an example, an unitialized PL/I variable declared as FIXED DECIMAL(3) would be represented as F30F00 (in hexadecimal). If this cell were assigned the value −45, its representation would change to F31045.

A *decimal-fixed-point* (dfx) cell has the second format in Figure 14.1. The size field defines the number of digits in the number. The fsize field indicates the number of digits to the right of an imaginary decimal point and must be less than or equal to size. The last field contains the

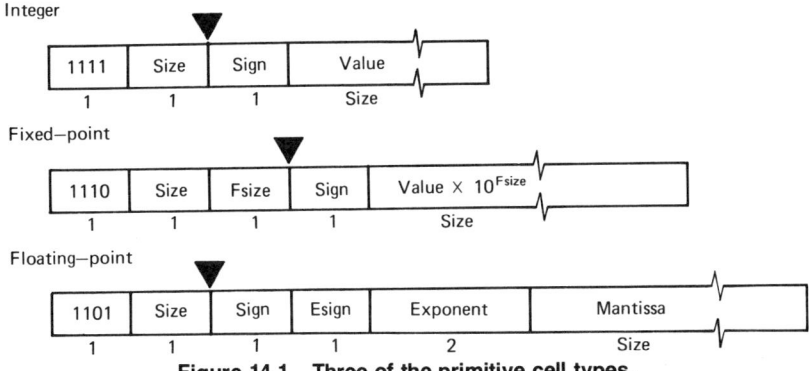

Figure 14.1 Three of the primitive cell types.

absolute value of the number times 10 to the power fsize. If the first value digit is 1111, the cell has the value "undefined." If a PL/I variable declared as FIXED DECIMAL(5,2) has the value 7.8, it would be represented as E52000780.

A *decimal floating-point* (dfl) cell has the third format in Figure 14.1. The second field defines the length of the mantissa, the third field describes the sign of the number, and the fourth field describes the sign of the exponent. The fifth field contains the absolute decimal value of the exponent (0–99). The last field contains the decimal mantissa. If the first 4 mantissa bits are 1111, the cell has the value "undefined." Operations on floating-point cells always normalize the mantissa (shift it so that no leading zeros occur, unless the cell's value is zero).

Note that all these representations allow for two representations of zero (+0 and −0). They are treated as identical; that is, comparing +0 and −0 for equality would result in a "true" condition.

The *boolean* (b) cell is illustrated in Figure 14.2. If the second field is 0000, the value is "false"; if the second field is 0001, the value is "true." If the second field is 1111, the value is "undefined."

A *character* (c) cell is shown in Figure 14.2. The second field contains the EBCDIC representation of a character. If it is 11111111, the value is "undefined."

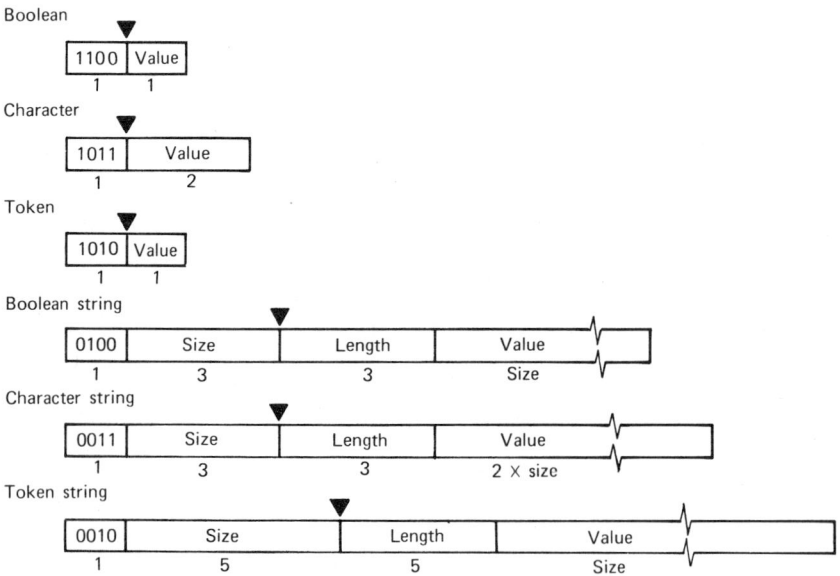

Figure 14.2 Six additional primitive cell types.

DATA TYPES

The value of a *token* (t) cell is any 4-bit quantity. The token data type is not used in most programs; it is intended for use by compilers and debugging tools, and the need for it will become obvious later. A token is the only cell type that cannot have the "undefined" value.

A *boolean string* (bst) cell is shown in Figure 14.2. The second field indicates the maximum number of booleans in the string (1–4095). The third field indicates the current number of booleans in the string (0–4095), allowing the string to shrink and grow dynamically (e.g., the VARYING string attribute in PL/I). The fourth field contains the actual string in which each element is represented as 0000, 0001, or 1111 (indicating an undefined element). If the length is zero, the entire string is interpreted as having the undefined value.

Another primitive cell is the *character string* (cst) cell. The fields have the meaning described previously. The fourth field contains the actual string in which each element is represented as 8 bits and the quantity 11111111 indicates an undefined element.

The last cell type illustrated in Figure 14.2 is the *token string* (tst) cell. The meaning of the fields is the same as that just described, but a token string has a maximum size and length of 1,048,575, and each element in the string is a 4-bit quantity. The token string is intended for use only by compilers and debugging tools.

The last primitive cell is a *pointer* (p); it has the form indicated in Figure 14.3. A pointer is a unique logical address of a cell or an *object* (module, activation record, or dynamically allocated storage area). Logical addresses are always assigned by the machine and can never be altered by a program (although a program is permitted to move the value of one pointer cell into another pointer cell). The representation of a logical address need be of no interest to a program, but seeing how the machine views it may clarify the purpose of a pointer. The first 36 bits of the logical address uniquely identify an object, and the last 44 bits identify a cell within the object. When an object (module, activation record, or dynamically allocated storage area) is created, the machine assigns it a unique 36-bit address and maintains a map to convert these addresses into physical addresses. Since the machine cannot create unique addresses forever using a finite-length address, once an object is destroyed its address will eventually be used for

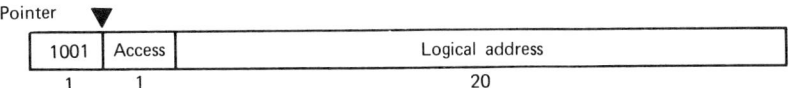

Figure 14.3 Format of the pointer cell.

another object, but not until approximately 80 days later (assuming that objects are created on the average of every 100 microseconds).

The second field (access) in a pointer cell represents an access or authority code to the object. When a logical address is created, the access code in the pointer is set to 1110, which denotes authority to read, write, and free the object. Other access codes are 1101 (read / write), 1100 (read-only), and 1011 (no access). An instruction is available to allow a program to alter the access code, but programs can only *lower* the access code in a pointer cell. If the access code is 1111, the pointer has the value "undefined."

Structure Cell Types

The only data type in this category is a structure cell. A structure describes a heterogeneous collection of other cells and corresponds to the concept of a structure in PL/I and COBOL. The properties that distinguish a structure from an array are that, in addition to the entire collection of elements having a name, in a structure each element also has a name and the elements in a structure can be different data types.

A *structure* (st) cell has a tag, but no content, component; its format is shown in Figure 14.4. The second field, a binary number from 1 to 255, specifies the number of cells included in this structure. The remaining fields specify the *cell addresses* of the cells that are included in this structure. A structure can include all cell types except a parameter cell. In other words, not only can a structure include heterogeneous primitive elements, but it can also include arrays and other structures.

The concept of a *cell address* is defined in a later section, but it is summarized briefly here. Each *module* (e.g., a PL/I external procedure) has an associated *address space* in which all cells reside that are accessible by the module. A cell address is simply the location of a cell within an address space. The size of a cell address can vary from module to module, and its size is denoted by N. As an example, the PL/I structure

```
DECLARE 1 PERSON,
   2 SALARY FIXED DECIMAL (7,2),
   2 NAME,
      3 LASTNAME CHARACTER (20),
      3 FIRSTNAME CHARACTER (12);
```

would result in five cells: two character-string cells, one decimal fixed-point cell, one structure cell pointing to the two strings, and

DATA TYPES

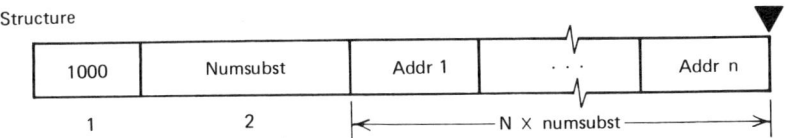

Figure 14.4 Format of the structure cell.

another structure cell pointing to the decimal fixed-point cell and the other structure cell.

Machine instructions can operate on structure cells as well as primitive and other cell types. For instance, a structure can be passed as an argument, or a structure can be moved into another structure (which causes the machine to locate and move the physical elements within the structure), providing that both structures have the same "structure."

A structure must be a tree as opposed to a lattice or network. For instance, a structure cannot contain multiple cell addresses to the same cell, and structure A cannot include structure B which includes structure A.

Nested Cell Types

The remaining three cell types are array, parameter, and relocatable. They are unique in that their tags are recursively defined; that is, their tags can include imbedded or nested tags.

An *array* (a) cell has the format shown in Figure 14.5. The second field is a binary number that specifies the number of dimensions (1–15). The third field is a nested tag; it is a tag describing the array elements. For instance, for an array of decimal integers, the third field would be 2

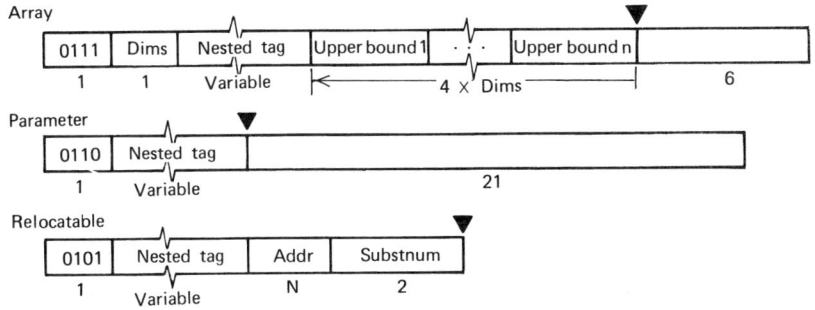

Figure 14.5 Format of the nested cell types.

tokens long and would contain 1111 followed by the integer size. For an array of character strings, this field would be 4 tokens long and would contain a tag of a character string. For an array of structures, this field would contain a variable-length structure tag. Valid element types in an array are all primitive cell types and structures.

Following the nested tag are one or more 4-token fields, one for each dimension. The binary value of each field indicates the upper bound of the array in the corresponding dimension. The product of these fields is equal to the total number of elements in the array (all dimensions have an implicit lower bound of 1).

Notice that the content component of an array cell is only 6 tokens long and obviously does not include enough space to store the elements. Since the machine will perfrom subscripting operations, the program need not know the physical location and organization of the elements. The 6-token field is used by the machine to identify the physical location of the elements. When an array is created (at the time the program is loaded, for "static arrays"; at the time an activation record is created, for "automatic arrays"; or at the time an array is explicitly dynamically allocated by the program), the machine allocates storage for the elements and places some internal address in the last field.

All subscripting is done by the machine, and many machine instructions function with entire arrays as well as array elements as operands.

To illustrate the meaning of the nested tag, a 3 by 3 array of FIXED DECIMAL(5,4) would be represented as

72E5400030003XXXXXX

A one-dimensional, 12-element array of boolean strings of size 10 would be represented as

71400A000CXXXXXX

A *parameter* (pm) cell has the format shown in Figure 14.5. Any variable in a module that is received as a parameter is defined by a parameter cell. The second field is a nested tag; it is a tag describing the attributes of the parameter and is used by the machine to check the correspondence between arguments and parameters. Valid tags are tags for all primitive cell types, structures, and arrays.

The last field is identical to the content component of a pointer cell (but a program cannot manipulate the access code of a parameter). This field is altered by the machine when a subroutine call occurs, and it is used to locate the argument.

DATA TYPES

A program uses a parameter cell as if it were a cell described by the nested tag. The only difference (which is transparent to the program) is that a reference to a parameter causes the machine to indirectly locate the storage via the last field. If the first token of the last field has the value 1111, the parameter has the value "undefined."

A *relocatable* (r) cell has the format shown in Figure 14.5. A relocatable cell represents a variable that is located relative to something else; that is, it is used as a "mask" that is placed over some storage area. Language data types that would be defined as relocatable cells are based variables, structures in an array, elements in a parameter structure, and all elements in a based structure or in a structure in an array.

The second field is a nested tag; it is a tag describing the attributes of the variable. Valid tags are tags for all primitive cell types, structures, and arrays.

The third field is a cell address that points to the cell from which this cell is relocatable. The cell address can only point to a pointer cell (for a variable based on a pointer), a structure (for a variable within a based structure, within a structure within an array, or within a parameter structure), or an array (for a structure that is an array element). The referenced cell can possibly be a parameter or relocatable.

If the relocatable cell points to a pointer or array cell, the fourth field has no meaning. If the relocatable cell points to a structure, the fourth field is a binary number (1–255) specifying that this relocatable cell corersponds to the *n*th element in the structure.

As was the case for a parameter, a program uses a relocatable cell as if it were not one; that is, it uses the cell as if its tag were the nested tag. The machine uses the last two fields to locate the appropriate storage location. If the relocatable cell represents a structure in an array or an element within a structure in an array, the program treats it as an array; that is, each machine instruction that refers to the relocatable cell must provide subscript information.

The uses of most of these 14 cell types are illustrated in the examples at the end of the chapter.

Note that 14 out of a possible 16 cell types have been defined, implying that only two more cell types could be added if the architecture is extended. This is not necessarily true; if the first 4 bits of a cell are 0000, this is intended to represent an "escape" code, meaning that the next 4 bits identify the cell type, allowing the machine to potentially have an unlimited number of cell types. A later section describes a feature of the architecture that allows it to have supplemental instruction sets; this escape code allows the supplemental instruction sets to define new cell types. For instance, if a FORTRAN-oriented supplemental instruction set is active, a cell beginning with the bits 00001111 might represent a

FORTRAN complex number (e.g., a numerical value with a real and an imaginary part).

THE MODULE

The principal storage object in the machine is the module. A module contains a sequence of machine instructions and a definition of the address space for those machine instructions.

A module corresponds to such programming language entities as PL/I external procedures, COBOL subprograms, and FORTRAN subroutine subprograms. The function of a compiler is to translate a high-level-language program into one or more modules. The format of these modules is shown in Figure 14.6. A module (represented as a token string) is defined to the machine via a LOAD-MODULE instruction, which causes the machine to make an internal copy of the module, perform certain storage management and initialization functions, and return a unique logical address of the module. A module consists of three variable-length components: the header, the address space, and the instruction space.

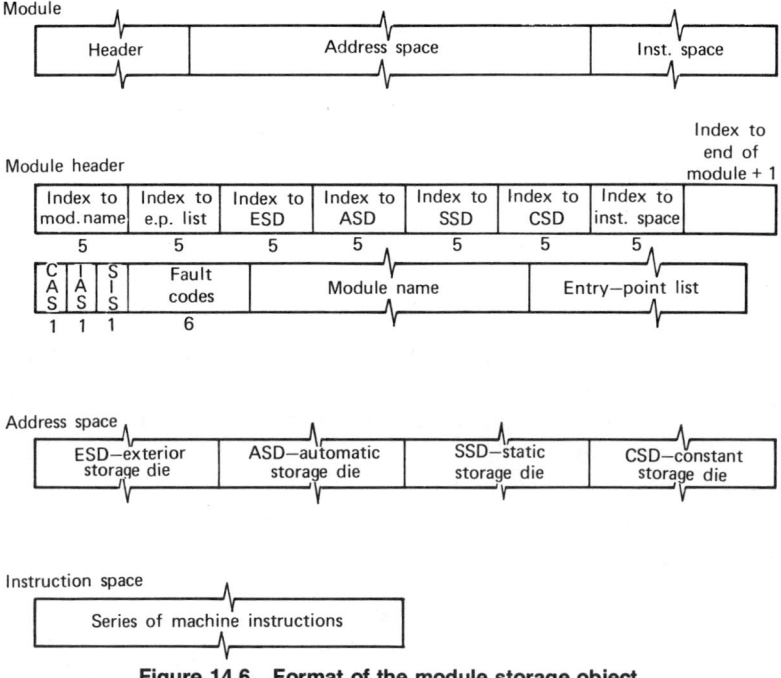

Figure 14.6 Format of the module storage object.

THE MODULE

The Header

The module header defines certain attributes of the module and sections of the other two components. The first eight fields in the header are 5-token fields containing the binary value of the index within the module of the beginning of a particular section of information (except for the eighth field, which indicates the end of the module). Since the index of a section is also used to indicate the end of the previous section, the seven sections must be contiguous. If a section is not present, its index field points to the start of the next section. For instance, if there is no exterior storage die section, its index field and the index field for the automatic storage die have the same value.

The next two 1-token fields (CAS and IAS) define the lengths of cell addresses and instruction addresses within this module. Each field can contain a binary value from 2 to 5, indicating 2–5 token addresses. Cell and instruction addresses are described in a later section on instruction formats.

Since the addressing space of a module is limited to only those cells defined in the module, it is desirable to limit the address-field sizes to the smallest size needed. That is, rather than defining fixed-length address fields within instructions, the size of an address field can vary from module to module. Cell addresses need only be large enough to address the cells within the module (the module's address space). Instruction addresses need only be large enough to address instructions within the instruction space. In other words, a module with only a few small cells (a small address space) needs only a tiny cell-address field; a module with more and bigger cells needs a larger cell address. The use of variable-size addresses is worthwhile because the physical size of the module can be reduced, the number of bits transmitted between the memory and the processor can be reduced, increasing the memory bandwidth, and arbitrary compromises concerning the upper bound of an address space can be avoided.

The next 1-token field (SIS—supplemental instruction set) in the header defines the language in which this module was written. The motivation for this field was the thought that the basic instruction set of the machine might be extended to provide additional instructions that are specialized toward a particular language. For instance, if this field is zero, operation code '0005' might be invalid. If the field is 1, operation '0005' might be a COBOL-oriented table search instruction; if the field is 2, operation '0005' might be a PL/I-oriented PICTURE editing instruction. If the field is 3, operation '0005' might be an instruction intended only for the operating system. This points out another motivation for such a feature: there is no need (or desire) to bother a COBOL compiler writer with information about instructions intended for the operating system. In fact, it is desirable to hide such instructions from those people and programs that have no direct use for them.

This "language" or supplemental instruction set field gives the machine the ability to vary part of its instruction set dynamically and gives system designer the ability to specialize and tailor the instruction set in a way that is transparent to existing programs. Although provision for this was made, no supplemental instruction sets currently exist, and the field should be set to zero.

The next 6-token field specifies the faults (machine-detected conditions) that this module wishes to handle. The meaning of this field is described in a later section on fault handling.

The next field is variable in length and contains the name of the module, using 2 tokens to represent each character.

The remaining part of the header is a list of the instruction addresses of the entry points in the module. The length of each field in this list is given by the previous field (IAS) defining the length of instruction addresses. A module with N source-language entry points has N+1 fields in this list. The first field is for entry point zero, a special entry point that is called when a fault occurs. If the module has no instructions to handle faults, this first field should be set to zero. The use of entry point zero is discussed in a later section on fault handling.

The Address Space

The second component of a module is its address space. The address space contains a series of cells defining the data that is accessible by the module. The index of the first token of a cell within the address space is known as its cell address. That is, the cell address of the first cell is 1; the cell address of the second cell is 1 plus the total length of the first cell, and so on.

Although the address space looks like one entity to the program, it is subdivided into four sections, as shown in Figure 14.6. These four sections are used by the machine for storage management, allocation, and protection purposes.

The *exterior storage die* holds cells that indirectly locate other cells (i.e., cells that have no value component). All (and only) relocatable and structure cells must be placed in the exterior storage die.

The *automatic storage die* holds all cells that are to be dynamically allocated space whenever the module is invoked. When the module is invoked, the machine allocates an activation record and copies the

THE MODULE

automatic storage die into the activation record. When the module's code refers to a cell in the automatic storage die, the machine automatically translates its cell address to a location within the activation record.

Note that the machine does a bit-by-bit copy of the automatic storage die into the activation record. This implies that the compiler can cause an automatic variable to have an initial value simply by putting the value in the variable's cell in the automatic storage die. If an automatic variable (or any other variable) has no defined initial value, the compiler is responsible for setting the cell to the "undefined" value. The only exception to this discussion is pointer and parameter cells; for reasons of security the machine always initializes them with the "undefined" value. All cell types except structure and relocatable may appear in the automatic storage die.

The *static storage die* holds all cells that are to be allocated once prior to execution (that is, at the time of the LOAD-MODULE instruction). If a static variable is to have an initial value, the value should be placed in its cell in the die. If not, the cell should be set to the "undefined" value. (All pointer cells are always initialized by the machine to the "undefined" value.) All cell types except structure and relocatable may appear in the static storage die.

The *constant storage die* has all the properties of the static storage die plus one additional property: the machine prohibits the alteration of cells within the constant storage die. The compiler should place all literal data within the constant storage die. If the programming language permits the declaration of variables as "read-only" or "constant," the cells corresponding to these variables should be placed in this die.

The Instruction Space

The last component of a module is its instruction space. The instruction space contains a series of machine instructions. Most machine instructions are represented in a variable number of tokens. The index of the first token of an instruction within the instruction space is known as its instruction address. The instruction address of the first instruction is 1; the instruction address of the second instruction is 1 plus the length of the first instruction, and so on.

Programming Note: Since array elements receive no space in the dies, it is not immediately obvious how a compiler would initialize an array. This suggestion is offered: If the array is automatic, the compiler must generate code (one or more MOVE instructions) to initialize the array at each entry point of the module. To initialize a static array, the compiler must give the module an extra entry point and generate code at this entry point to initialize the array. After the LOAD-MODULE has been performed, this special entry point should be called to initialize the static array. Since cells in the constant storage die cannot be altered, there is no known way to initialize an array in the constant storage die.

INSTRUCTION FORMATS AND ADDRESSING

A machine instruction consists of an operation code followed by one or more address fields. Some instructions have only one address field; others have two, others have three, and certain instructions have a variable number of address fields.

Operation Codes

The first field of each instruction is the operation code. Rather than using a single-length field for operation codes, a frequency-based encoding was done. That is, the operation-code field for the 15 most frequent instructions is 1 token long, the field is 2 tokens long for the second most frequent set of 15 instructions, and so on. The motivation for a frequency-based encoding, the rationale for choosing this particular encoding, and the selection of the operation codes are discussed in Chapter 17.

Address Fields

There are seven types of address fields, grouped into three categories: operand addresses, instruction addresses, and immediate fields.

An *operand address* references an operand in the address space. There are four types of operand addresses:

1. *Cell Address.* A cell address is an N-token binary field that refers to a cell in the address space (N is the value of the cell-address-size field in the module header). For instance, if N (CAS) has the value 2, the operand address 1A refers to the cell beginning at the 26th token in the module's address space.
2. *Literal.* A literal field consists of N tokens of zeros followed by one token having the value zero, one, ..., or nine. A literal field is assumed to be a one-digit positive decimal integer. As an example, if N (CAS) has the value 2, the operand address 004 is a literal of value +4.
3. *Array Element Address.* An array element address consists of D+1

FAULT HANDLING

subfields. The first subfield is a cell address of an array (or a relocatable cell defining an array element) having D dimensions. The next D subfields are cell addresses, literals, or array element addresses specifying the values of the subscripts (the values must be decimal integers). For example, if array cell A has the index (in hexadecimal) of 20 in the address space, if a variable I has the index 3C, and if N is 2, then the operand address for A(4,I) is 200043C. If N were 3, the operand address would be 020000403C.

4. *Array Address.* An array address refers to an entire array. Array addressing is identical to array element addressing, except that all the subscript subfields are specified as *. The * is represented by a literal field with the value F (1111). Hency array A is addressed by 2000F00F. This addressing scheme allows for the possibility of addressing array cross-sections (e.g., the PL/I expression A(*,I) would produce the address 2000F3C), although this is not currently part of the architecture.

Unless otherwise mentioned, any of these four forms can be used as operand addresses in instructions. One general exception is that a literal cannot be used as a *target operand*. An *operand* is the data referred to by an operand address (perhaps indirectly through a relocatable, structure, or parameter cell); a target operand is an operand in which an instruction stores a result.

The second category of address fields is an *instruction address*. An instruction address is an M-token field that refers to an instruction in the instruction space (M is the value of the instruction-address-size field in the module header).

The last category of address fields is an *immediate field*. An immediate field is a 1- or 2-token field containing not an address but some value that is used directly by the instruction. Since immediate fields have specialized purposes and are only used in a few instructions, definition of the immediate fields is deferred to the definitions of these instructions.

To avoid repetitive lengthy descriptions in the definitions of the instructions, certain groups of operands are named. *Arithmetic operands* include decimal integers (and literals), decimal fixed-point cells, and decimal floating-point cells. *String operands* include boolean strings, character strings, and token strings.

FAULT HANDLING

Since the major objective of this machine is to prevent and/or detect certain classes of programming errors, the methods by which the machine detects and reports errors are of special importance. This section defines the conditions (called *faults*) detected by the machine, the information that the machine presents to the program when a fault occurs, and how the program and machine can interact to handle faults.

Fault Descriptions

The following descriptions define the types of faults detected by the machine and the situations in which they arise.

An *invalid operation* (type 1) fault occurs when the machine fetches an instruction but its operation code is invalid, or when the end of the instruction space is encountered during the fetching of an instruction.

An *addressing* (type 2) fault occurs when a cell address is being used but it falls beyond the module's address space or resides in an incorrect storage die, or when an array subscript is not a decimal integer.

An *unknown data format* (type 3) fault occurs when the machine references a cell that has an unrecognizable format or value.

A *protection* (type 4) fault occurs when (1) the program attempts to free, write to, or read from a cell that is located through a pointer or parameter cell, but the pointer or parameter does not have the appropriate access code, (2) the program attempts to explicitly free storage that resides within an activation record or module, and (3) the program attempts to alter a cell within a constant storage die.

An *invalid pointer* (type 5) fault occurs when the program uses a pointer or parameter cell but the logical address in the pointer is unknown to the machine (implying that the storage referred to by the pointer has been previously freed).

A *bounds-exceeded* (type 6) fault occurs when the program refers to an array element using a subscript that is beyond the bounds of the corresponding dimension, or when a program refers to a string element that is beyond the current length of the string.

An *invalid operand type* (type 7) fault occurs when the type of an operand does not match the valid operand type(s) in the instruction specification.

An *undefined operand* (type 8) fault occurs when the machine attempts to use the value of an operand, but (1) the operand, or (2) a pointer or (3) parameter cell used to locate the operand, has the value "undefined." This fault does not occur for condition 1 in the DEFINED instruction, which is an explicit test to determine if an operand has an undefined value.

An *incompatible operands* (type 9) fault occurs when two or more operands of an instruction are incompatible. The conditions of operand compatibility are defined in the specifications of the instructions. This

FAULT HANDLING

fault can also occur in an ACTIVATE or LOCAL-ACTIVATE instruction when the type of a parameter cell is incompatible with the type of the corresponding argument cell. The fault also occurs when the attributes of a "locator" cell (e.g., relocatable cell) do not match the attributes of the indirectly located data.

An *overflow* (type 10) fault occurs when the target operand in an instruction is too small to hold the value produced by the instruction. For arithmetic operands this occurs when loss of high-order nonzero digits would occur or when the exponent of a floating-point result is greater than 99. For string operands this occurs when the size of the target string is too small to hold the value produced by the instruction.

An *underflow* (type 11) fault occurs when the floating-point result of an instruction has an exponent of less than -99.

A *divide* (type 12) fault occurs when division by zero is attempted.

An *address loop* (type 13) fault occurs when the machine encounters a loop while searching a chain of addresses. This can occur when the machine is tracing through structure cells looking for the atomic cells, or when following a pointer chain with the CHAIN instruction.

An *invalid module* (type 14) fault occurs during a LOAD-MODULE instruction when the machine discovers a format error in the module being loaded.

An *invalid transfer* (type 15) fault occurs for various reasons in an instruction that transfers control flow. The most common situation is attempting to branch beyond the instruction space of the module.

An *invalid parameter count* (type 16) fault occurs in an ACTIVATE or LOCAL-ACTIVATE instruction when the number of parameters specified does not equal the number of arguments transmitted.

A *conversion* (type 17) fault occurs during the CONVERT instruction when the operands do not match the conversion rules listed in the specification of the CONVERT instruction.

A *branch-trace* (type 18) fault occurs during the execution of a BRANCH or BRANCH-FALSE instruction if branch tracing has been enabled.

A *call-trace* (type 19) fault occurs during the execution of a CALL or LCALL instruction if call tracing has been enabled.

A *signal* (type 20) fault occurs when the SIGNAL instruction is executed.

Entry Point Zero

Each fault type has an associated number, given in the previous section. These numbers also correspond to a bit position in the fault-code field in the module header. For example, fault type 1 (invalid operation) corresponds to the first bit in the fault-code field. If a bit is set to 1 in the fault-code field in a module, this indicates that this module desires to handle the associated fault.

When a fault occurs, the machine attempts to call entry point zero of the current module. Entry point zero will be called if it exists (i.e., the first entry in the entry-point list in the module header is nonzero) and if the fault is enabled (the corresponding bit in the fault-code field is 1). If these two conditions are not met, the machine attempts to call another entry point zero by searching backwards through the stack of active modules until a module is found with this fault enabled. It is anticipated that the first module invoked in each program is a special module generated by the compiler or operating system that has all faults enabled.

When an entry point zero (hereafter called a *fault handler*) is invoked, it is called by the machine as an internal procedure. Therefore, the fault handler has addressability to the address space in the module in which the fault handler resides. The machine also passes these five arguments to the fault handler:

1. A decimal integer of size 2 containing the fault type.
2. The pointer of the module in which the fault arose.
3. A token string of size 5 containing the instruction address of the instruction causing the fault.
4. A token string of size 5.
5. A pointer.

The fault handler is given read-only access to the arguments.

The last two arguments are used to convey additional diagnostic information about the fault. Their meaning depends on the type of fault.

For the invalid-module fault, argument 4 is an error code describing the error in the module. For the signal fault, argument 4 is the 2-token immediate field in the SIGNAL instruction. For the branch-trace fault, argument 4 has the value 1 (length = 1) if the branch will be taken, and the value 0 if the branch will not be taken.

For all faults except call-trace, argument 5 has the undefined value. For the call-trace fault, arguments 4 and 5 have this meaning: If the faulting instruction was CALL, argument 4 is the entry-point number and argument 5 is a pointer to the called module. If the faulting instruction was LCALL, argument 4 is the instruction address of the called routine and argument 5 is undefined. For all faults not mentioned in the last two paragraphs, argument 4 is undefined (zero length).

INSTRUCTION SUMMARY

Machine instructions are available to allow a module to dynamically enable or disable specific faults and to allow a fault handler to resume execution at the instruction following the faulting instruction, retry the faulting instruction, or transfer the fault to a higher fault handler.

A fault in a fault handler is treated like any other fault situation. The only difference is that, to prevent a faulting fault handler from entering endless recursion, the search for an applicable fault handler starts with the module that called the module containing the faulting fault handler.

Program State After a Fault

A key consideration in fault handling is the state in which the machine leaves the program when a fault occurs. In most cases the faulting instruction does not affect the state of the program. A fault handler should terminate with one of four instructions: LOCAL-RETURN, which terminates the fault handler and resumes execution of the faulting instruction, CONTINUE, which terminates the fault handler and resumes execution at the instruction following the faulting instruction, RETURN, which deletes the activation records for this module and all later modules and returns control to the module that previously called this module, and TRANSFER-FAULT, which terminates the fault handler and causes the machine to search for and call a higher fault handler. The exceptions to these general rules are discussed below.

1. Issuing the LRETURN instruction after a branch-trace or call-trace instruction causes the faulting instruction to resume as if the fault had not occurred.
2. Faults that occur during the processing of a structure or array result in the fields or elements processed before the fault taking on their new values, but all remaining fields or elements remain unchanged.

INSTRUCTION SUMMARY

This section summarizes the instructions of the machine. Chapter 15 describes each instruction in greater detail.

General Instructions

The three general instructions are MOVE, CONVERT, and UNDEFINE. Operands of the three instructions may be single scalar cells, arrays, strings, and structures. MOVE is used to transfer the value of one operand to another. CONVERT performs the same function as MOVE, but it also performs an explicit data conversion. For instance, if one used a MOVE instruction to move a character value into a decimal integer, the operation would fail and an incompatible-operands fault would occur. If one used a CONVERT instruction, the operation would succeed; the character value would be converted into an integer according to a set of predefined rules.

The UNDEFINE instruction is used to set the value of an operand to undefined.

Arithmetic Instructions

The arithmetic instructions include ADD, SUBTRACT, MULTIPLY, DIVIDE, MODULO, ABSOLUTE, COMPLEMENT (unary minus), and POWER (compute X to the Yth power). The ADD, SUBTRACT, MULTIPLY, DIVIDE, MODULO, and POWER instructions have two operands; the result is stored in the first operand. ABSOLUTE and COMPLEMENT have one operand. The operands must be arithmetic scalars or arrays.

Comparison Instructions

The EQUAL, NOT-EQUAL, LESS-THAN, GREATER-THAN, LESS-THAN-OR-EQUAL, and GREATER-THAN-OR-EQUAL instructions have three operands; the values of the second and third operands are compared, and the boolean result is placed in the first operand. In general, the two comparison operands may be any cell types (e.g., pointer, character string, array, structure).

The last comparison instruction is DEFINED. This instruction tests an operand to determine if its value is defined and places the boolean result in the other operand.

Boolean Instructions

The boolean instructions are AND and OR, which have two operands, and NOT, which has one operand. The operands may be boolean, boolean strings, or arrays of booleans or boolean strings.

String Instructions

Although many of the machine instructions can have string operands, the string instructions work exclusively with string operands. The operands may be boolean, character, or token strings.

The CONCATENATE instruction appends the value of one operand

INSTRUCTION SUMMARY

to the end of the other operand. The MOVESUBSTRING instruction overlays a substring in one operand onto a substring in the other operand. The INDEX instruction searches a string for a designated substring. The LENGTH instruction returns the current length of a string.

Control Instructions

The control instructions are associated with transfers of execution flow. The CALL, ACTIVATE, and RETURN instructions are associated with calls to modules; the LOCAL-CALL, LOCAL-ACTIVATE, and LOCAL-RETURN instructions are associated with calls to local subroutines within a module; the BRANCH and BRANCH-FALSE instructions alter execution flow within a module.

The CALL instruction specifies the module being called, the entry-point number in the module, and a list of arguments. A subset of these arguments may be designated as read-only, implying that the called module may not alter nor free them. CALL allocates the storage specified in the automatic storage die of the called module and branches to the specified entry point. The first instruction at each entry point must be an ACTIVATE instruction, which specifies a list of parameters. The instruction checks the compatibility of the arguments and parameters and initializes the parameters (the transmission method is by-reference). The RETURN instruction frees the automatic storage and transfers control to the module that called this module.

The LCALL instruction specifies an instruction address of a local procedure and a list of arguments. The first instruction of a local procedure must be LACT (LOCAL-ACTIVATE). LACT specifies a list of parameters and causes the compatibility of the arguments and parameters to be checked and the parameters to be initialized. LCALL does not allocate any automatic storage, which means that the machine provides only minimal support of local procedures. If storage allocation and scope-of-name rules are necessary, they are the compilers' responsibility. The LRETURN instruction transfers control back to the instruction following the last LCALL instruction.

The BRANCH instruction transfers control to a specified instruction address. The BRANCH-FALSE instruction transfers control to a specified instruction address if a boolean operand has a zero (false) value or to the next sequential instruction if the operand is 1 (true).

Addressing Instructions

This group of instructions is associated with the manipulation of pointers and storage objects. The CREATE-POINTER instruction produces a pointer to a specified operand. The CHANGE-ACCESS instruction is provided to lower (further restrict) the access code in a pointer. The ALLOCATE instruction is used to allocate storage space dynamically, and the FREE instruction is used to free an object dynamically (i.e., a module or a dynamically allocated storage space).

The CHAIN instruction follows a chain of pointers and returns the value of the last one. The LOAD-MODULE instruction defines a module to the machine and returns a pointer to it. The LINK instruction is used to assign a value to a pointer cell in a loaded module (LINK is used for "link-editing" functions).

Debugging Instructions

The last set of instructions is associated with debugging and fault-handling functions. The ENABLE and DISABLE instructions provide the program with a way to dynamically enable or disable faults designated for the module's fault handler. The SIGNAL instruction is used to explicitly trigger a fault and enter a fault handler.

The CONTINUE instruction provides a fault handler with the ability to resume execution of the faulting module at the instruction following the faulting instruction. (LRETURN is used to resume execution at the faulting instruction.) If a fault handler determines that a fault should be transferred to a "higher" fault handler, the TRANSFER-FAULT instruction is used.

The DISPLAY-TAG and DISPLAY-CONTENTS instructions are intended for debugging operations. Given a cell address and a pointer to a module, the instructions will place either the tag or the content of the referenced cell into a token string.

The TRACE and NOTRACE instructions are used for monitoring execution flow. The TRACE instruction enables a trace of branch instructions, call instructions, or both in a specified module or set of modules, and the NOTRACE instruction disables the same. If a branch trace is enabled for a module, all BRANCH and BRANCH-FALSE instructions generate a branch-trace fault.

A ONE-MODULE EXAMPLE

To illustrate the SWARD architecture, the compilation and execution of several programs are analyzed. The first program is a PL/I version of the second Student-PL program in Chapter 6, which was also the program used in Chapter 9 to illustrate the SYMBOL architecture. The PL/I version of the program is

A ONE-MODULE EXAMPLE

```
 1  (SUBSCRIPTRANGE): XYZ: PROCEDURE;
 2  DECLARE A(4) FLOAT BINARY(21);
 3  DECLARE (B,C) FIXED BINARY(31);
 4  B=4;
 5  A=1;
 6  C=B+B*B;
 7  CALL ZZZ(B);
 8  A(B)=C;
 9    ZZZ: PROCEDURE(M);
10    DECLARE M FIXED BINARY(31);
11    C=M+M;
12    END;
13  END;
```

Our goal here is to build a token string containing a module representing this program and to consider how this module would be executed. To begin, we could consider the construction of the module header, but most of the fields in the header cannot be given values until the remainder of the module is generated. Hence the construction of the module header is deferred, but we assume that the module is small enough (in terms of both data and instructions) to allow CAS and IAS to be set to 2 (i.e., 2-token cell and instruction addresses).

The first step is to consider the module's address space. All the variables in the above program are to be represented in the automatic storage die. Since this PL/I program uses binary arithmetic but the machine supports only decimal arithmetic, the decimal-equivalent precisions must be determined. A floating-point binary number of 21 mantissa bits is equivalent to a decimal mantissa of seven digits, and a binary integer of 31 bits is equivalent to a decimal integer of 10 digits.

The module's address space is defined in Figure 14.7. Since the exterior storage die is empty, the indexes shown are also the cell addresses. Since the variables have no initial values in the source pro-

Index		Comments
01	71D70004000000	Floating-point array A
0F	FA0F000000000	Integer B
1C	FA0F000000000	Integer C
29	6FAF00000000000000000000	Parameter integer M

Figure 14.7 Address space of module XYZ.

gram, they are given the undefined value. The reader should analyze each cell's representation before reading further.

The next consideration is the object code needed to represent the program. For ease of representation, the object code is represented in a hypothetical assembly language. The machine instruction and its index (instruction address) in the instruction space are shown next to each assembly-language statement.

The first instruction in a module must be an ACTIVATE instruction. This instruction has a variable number of address fields. The first field is an immediate field specifying the number of parameters. The remaining fields are operand addresses of the parameters. ACTIVATE initializes the parameters and checks the compatibility of each parameter and the corresponding argument. For this program, the instruction generated is

ACT 0   (01: 0900)

The instruction is represented as 0900; the op-code is 09, and the instruction address is 01. The immediate field 00 indicates that zero parameters are received. If parameters were expected, the instruction would also contain cell addresses of the parameters.

After the program has been compiled and loaded, this module (XYZ) is called by some other module (e.g., the operating system). At that time, an activation record is built, the automatic storage die is copied into the activation record, and the storage for the elements of array A is allocated. However, the instructions in the XYZ module address cells as if they physically reside in the module. If a referenced cell is in the automatic storage die, the machine automatically translates the reference to a location in the activation record.

The next statement is B=4. The MOVE instruction is used, and since the assigned datum (4) is an integer constant of value less than 10, a literal is used in the instruction. The instruction generated is

MOVE B,'4'   (05: 10F004)

The MOVE instruction, a frequently used instruction, has a 1-token op-code (1). 0F is the cell address of B, and 004 is the literal. The cell representing B would be

FA00000000004

after the execution of this instruction.

A ONE-MODULE EXAMPLE

Statement 5 (A=1) is similar and would be represented as

MOVE A,'1'   (0B: 10100F001)

In the machine instruction, 0100F is an array address of A (01 is the cell address and 00F is the * subscript). Note that a scalar is being moved into an array and that the scalar is an integer but the array is floating-point. The machine recognizes both situations and assigns each element of A the floating-point representation of 1.

The code generated for statement 6 (C=B+B*B) is

```
MOVE   C, B   (14: 11C0F)
MULT   C, B   (19: 51C0F)
ADD    C, B   (1E: 31C0F)
```

Each of these instructions checks to ensure that the source operand has a defined value, that both operands are arithmetic, and that the target location is large enough to hold the result.

Statement 7 is the CALL statement. The instruction generated is

LCALL ZZZ,1,0,B   (23: 0B??01000F)

The op-code is 0B. The next field is the instruction address of the called procedure within this module, but its address has not yet been determined. The next field is a 2-token immediate field indicating the number of arguments. The fourth field is a 2-token immediate field indicating that the called procedure is to be restricted to read-only access to the first N arguments. Since the PL/I language does not provide for this, the field is set to zero. The last fields (one in this case) are the operand addresses of the arguments. The LCALL (local-call) instruction suspends execution at the current location and transfers control to the specified instruction address.

Statement 8 illustrates the array-subscripting aspect of the SWARD architecture. The instruction generated is

MOVE A(B),C   (2D: 1010F1C)

The 010F is an array-element address; 01 is the cell address of A, and 0F is the cell address of the single subscript to this one-dimensional array. Note that if the current value of B were beyond the defined bound of A, a bounds-exceeded fault would be generated.

Since statement 9 is the beginning of an internal procedure, we must branch around it; thus the instruction

    B   ?   (34: E??)

is generated. (Since we have not yet completed the compilation, the branch address is unknown.) The next instruction is

    LACT 1,M   (37: 0C0129)

The local-activate instruction is similar to the earlier activate instruction. One parameter is specified, implying that the instruction will check the compatibility of the parameter and the argument and initialize the parameter with a logical address pointing to the argument.

For statement 11, the instructions

    MOVE   C,M   (3D: 11C29)
    ADD   C,M   (42: 31C29)

are generated. The machine recognizes that the source operands are parameters; it automatically traces back to the argument and uses the argument's value as the source value.

The remainder of the module is

    LRETURN   (47: 0D)
    RETURN   (49: 0A)

The LRETURN (local-return) instruction transfers control back to the instruction following the last-executed LCALL instruction (to instruction address 2D in this case). The RETURN instruction is now seen to be the target of the earlier branch instruction; it terminates this invocation of the module and returns control to the module that called this module.

The completed form of the module is shown in Figure 14.8. Rather than illustrating the module as a contiguous set of tokens, items of interest (e.g., individual cells and instructions) are illustrated on separate lines. The first and second columns are not part of the module; they indicate, respectively, the index in the module of the first token on the line and the index in the address space or instruction space of the first token on the line. Each line is also described by a comment that, of course, is not a part of the module.

A ONE-MODULE EXAMPLE

```
Indexes                                             Comments
001        00032000380003C0003C                     Header
015        0007C0007C0007C000C6                     Header
029        220000000                                Header
032        E7E8E9                                   Module name
038        0001                                     Entry-point list
03C    01  71D70004000000                           A (floating-pt array)
       0F  FA0F000000000                            B (integer)
       1C  FA0F000000000                            C (integer)
       29  6FAF00000000000000000000                 M (integer parameter)
07C    01  0900                              ACT    0
       05  10F004                            MOVE   B,'4'
       0B  10100F001                         MOVE   A,'1'
       14  11C0F                             MOVE   C,B
       19  51C0F                             MULT   C,B
       1E  31C0F                             ADD    C,B
       23  0B3701000F                        LCALL  ZZZ,1,0,B
       2D  1010F1C                           MOVE   A(B),C
       34  E49                               B      %A
       37  0C0129                      ZZZ:  LACT   1,M
       3D  11C29                             MOVE   C,M
       42  31C29                             ADD    C,M
       47  0D                                LRETURN
0C4    49  0A                          %A:   RETURN
```

Figure 14.8 Compiled module XYZ.

Table 14.1 continues the case-study comparison by indicating the attributes of the equivalent Student-PL program on the Student-PL machine, the equivalent SPL program on the SYMBOL machine, and this PL/I program on the S/370 and SWARD machines.

The most useful way to compare these architectures is an analysis of the number of bits that must move between the memory and the processor to execute the program. However, lacking this information, the data in Table 14.1 can be used as an extremely gross indication of relative performance, and they are interesting in that the SWARD machine appears to exceed even the two earlier non-von Neumann architectures.

Table 14.1 Relative Program Sizes

Machine	Instructions Executed	Object-Code Size (bytes)	Total Program Size (bytes)
Student-PL	62	96	69
SYMBOL	81	324	396
S/370	117[a]	392	5536
SWARD	14	37	98.5

[a] Plus approximately 90,000 instructions at the beginning of the program to set up storage for the call/return mechanism.

A TWO-MODULE EXAMPLE

The second example is a PL/I program consisting of two external procedures. The two procedures are shown in Figures 14.9 and 14.10. The example illustrates many of the attributes of the architecture in that it includes character strings, structures, arrays, based variables, pointer variables, a structure containing an array of structures, a module call, and both static and automatic storage.

The object module for procedure TESTEST is shown in Figure 14.11. The module was small enough to be compiled with 2-token cell and instruction addresses. The exterior storage die contains cells representing TAB, HEAD, N, T, and A. TAB and HEAD, PL/I structures, are represented as structure cells. For instance, the cell for TAB, 8020873, indicates that TAB has two substructures at cell addresses 08 and 73.

Variables N, T, and A have no fixed location because they are elements of a structure within an array; they are represented as relocatable cells. Looking at the representation of N (530087301), 3008 is the nested tag describing N as a character string of size 8, 73 is the address of the cell from which N is relocatable (array B), and 01 indicates that N is the first structure element.

The automatic storage die contains the cells for ESTAB, UNNAME, and CODE. When the activation record is created, these variables will be initialized to undefined, XXXXXXXX, and 16, respectively.

The variables TAG, SIZE, B, and MATCHES are represented by cells

```
TESTEST: PROCEDURE OPTIONS(MAIN);
DECLARE ESTAB POINTER;
DECLARE 1 TAB STATIC,
          2 HEAD,
            3 TAG CHAR(4) INIT('ESTB'),
            3 SIZE BINARY FIXED(15) INIT(0),
          2 B(7),
            3 N CHAR(8),
            3 T CHAR(2),
            3 A POINTER;
DECLARE UNNAME CHAR(8) INIT('XXXXXXXX'),
        CODE   FIXED BINARY(15) INIT(16);
ESTAB=ADDR(TAB);
SIZE=1;
B(1).N='A';
B(1).T='EP';
B(1).A=ADDR(UNNAME);
CALL MATCHES(ESTAB,UNNAME,CODE);
END;
```

Figure 14.9 PL/I procedure TESTEST.

```
MATCHES: PROCEDURE(ESTAB,UNRESNAME,MATCHCODE);
DECLARE ESTAB POINTER;
DECLARE 1 TABLE BASED(ESTAB),
          2 HEADER,
            3 TAG CHAR(4),
            3 SIZE BINARY FIXED(15),
          2 BODY(2000),
            3 NAME CHAR(8),
            3 TYPE CHAR(2),
            3 ADDRESS POINTER;
DECLARE
        MODULE  CHAR(2) STATIC INIT('MD'),
        ENTRYPT CHAR(2) STATIC INIT('EP'),
        EXTREF  CHAR(2) STATIC INIT('ER');
DECLARE NULL BUILTIN;
DECLARE MATCHCODE FIXED BINARY(15);
DECLARE UNRESNAME CHAR(8);
DECLARE
        I FIXED BINARY(15);
        J FIXED BINARY(15);

MATCHCODE=2;
IF(ESTAB¬=NULL)
  THEN
    IF((TAG='ESTB')&(SIZE>0)&(SIZE¬>2000))
      THEN
        DO;
          MATCHCODE=0;
          DO I=1 TO SIZE WHILE(MATCHCODE=0);
            IF(BODY(I).ADDRESS=NULL)
              THEN DO;
                  MATCHCODE=1;
                  DO J=1 TO SIZE WHILE(MATCHCODE=1);
                    IF((BODY(I).NAME=BODY(J).NAME)&
                       ((BODY(J).TYPE=MODULE)|
                        (BODY(J).TYPE=ENTRYPT)))
                      THEN DO;
                          MATCHCODE=0;
                          BODY(I).ADDRESS=BODY(J).ADDRESS;
                          END;
                      ELSE;
                    END;
                  IF(MATCHCODE=1) THEN UNRESNAME=BODY(I).NAME;
                                  ELSE;
                  END;
              ELSE;
            END;
          END;
      ELSE;
  ELSE;
END;
```

Figure 14.10 PL/I procedure MATCHES.

```
Indexes                                    Comments
01       0003200040000440006A              Header
15       0009F000E1000EF0012B              Header
29       220000000                         Header
32       E3C5E2E3C5E2E3                    Module name
40       0001                              Entry point list
44   01  8020873                           TAB (structure)
     08  8025C6B                           HEAD (structure)
     0F  530087301                         N (relocatable char. string)
     18  530027302                         T (relocatable char. string)
     21  597303                            A (relocatable pointer)
6A   27  9F0000000000
         0000000000                        ESTAB (pointer)
     3D  3008008E7E7E7
         E7E7E7E7E7                        UNNAME (char. string)
     54  F5000016                          CODE (dec. integer)
9F   5C  3004004C5E2E3C2                   TAG (char. string)
     6B  F5000000                          SIZE (dec. integer)
     73  718030F1821
         C007000000                        B (array of structures)
     88  9F0000000000
         0000000000                        MATCHES (pointer)
E1   9E  BC1                               'A' (char.)
     A1  3002002C5D7                       'EP' (char. string)
EF   01  0900                    ACT       0
         0E2701                  CPTR      ESTAB,TAB
         16B001                  MOVE      SIZE,'1'
         10F0019E                MOVE      B.N('1'),'A'
         118001A1                MOVE      B.T('1'),'EP'
         0E210013D               CPTR      B.A('1'),UNNAME
         D880010300273D54 CALL             MATCHES,'1',3,0,ESTAB,
                                           UNNAME,CODE
129      0A                      RETURN
```

Figure 14.11 Compiled module TESTEST.

in the static storage die. Taking a closer look at the cell representing B, since B is an array of structures, it contains the nested tag 8030F1821. Since B is in the static storage die, the machine would allocate storage for the array when the module is loaded and put the address of this storage in the last 6 tokens of the cell.

The procedure contains two constants; these are represented in the constant storage die. Recall that this storage is the same as static storage, except that it is read-only. For instance, if the module contained the statement

CALL SUB ('A', CODE);

and procedure SUB attempted to store into the first parameter, a protection fault would occur.

Looking at the object code, the CPTR (create pointer) instruction

A TWO-MODULE EXAMPLE

assigns the logical address of TAB to ESTAB. In this case the logical address of TAB consists of the 36-bit unique identification of this module (since TAB is in the static storage die and therefore is physically contained in the module storage object) and the 44-bit cell address of TAB. (To a program, a logical address is an 80-bit name. The above statement was included to indicate how the machine forms logical addresses.)

To understand the use of the relocatable cells, examine the second MOVE instruction (10F0019E). 0F is the cell address of N, but the machine recognizes that N is a relocatable structure element in an array; thus it expects the operand address to be an array address (referencing all occurrences of N) or an array-element address (referencing one occurrence of N). In this case the operand address is 0F001, indicating the first occurrence (001 is a literal subscript). Examine Figure 14.11 to convince yourself that the machine has enough information to locate the storage into which the value 'A' is moved.

The CALL instruction (op-code = D) creates an activation record for the module referenced by the pointer at address 88 and transfers control to entry point 1 of the module. The pointer at location 88 would have been given a value during the linkage-editing process. The CALL instruction passes three arguments (the cells at addresses 27, 3D, and 54).

Figure 14.12 defines the module header and address space for the procedure MATCHES in Figure 14.10. Note that the instruction-address-size field is set to 3. The module was originally compiled with an IAS of 2, but this proved to be insufficient to address the entire instruction space for this module.

The structure TABLE is a based variable, based on the parameter ESTAB; thus the eight identifiers in the structure are represented by relocatable cells. The cell representing TABLE is relocatable with respect to the cell at location 55, the cell representing the first parameter. The reader is advised to trace the relationships in Figure 14.12 to see how the structure TABLE is represented.

Note the three boolean cells labelled $B1, $B2, and $B3. These are temporary variables generated by the compiler to keep track of the conditions tested in some of the multiple-condition IF statements. Also note that all variables that do not have a designated initial value are set to the undefined value by the compiler.

The instruction space of the module is illustrated in Figure 14.13. The ACT (activate) instruction checks the three parameters for consistency with the three arguments and initializes the three parameters with the logical addresses of the arguments. The DEF (DEFINED) in-

```
Indexes                                    Comments
001          00032000400004600009A         Header
015          000F90011A001300023B          Header
029          230000000                     Header
032          D4C1E3C3C8C5E2                Module name
040          000001                        Entry point list
046    01    58020D295500                  TABLE (relocatable structure)
       0D    580219220101                  HEADER (relocatable structure)
       19    530040D01                     TAG (relocatable char. string)
       22    5F50D02                       SIZE (relocatable dec. integer)
       29    5718033D464F
             07D00102    BODY              (relocatable array of structures)
       3D    530082901                     NAME (relocatable char. string)
       46    530022902                     TYPE (relocatable char. string)
       4F    592903                        ADDRESS (relocatable pointer)
09A    55    69F00000000000
             000000000                     ESTAB (parameter pointer)
       6C    6F5F0000000000
             000000000                     MATCHCODE (parameter dec. integer)
       84    63008F0000000000
             000000000                     UNRESNAME (parameter char. string)
       9E    F50F0000                      I (dec. integer)
       A6    F50F0000                      J (dec. integer)
       AE    CF                            $B1 (boolean)
       B0    CF                            $B2 (boolean)
       B2    CF                            $B3 (boolean)
0F9    B4    3002002D4C4                   MODULE (char. string)
       BF    3002002C5D7                   ENTRYPT (char. string)
       CA    3002002C5D9                   EXTREF (char. string)
11A    D5    3004004C5E2E3C2               'ESTB' (char. string)
       E4    F402000                       '2000' (dec. integer)
```

Figure 14.12 Module header and address space for MATCHES.

struction (op-code = 004) corresponds to the condition ES-TAB≠NULL is the source program. If ESTAB (actually, the argument pointed to by the cell representing ESTAB) has a defined value, the cell at location AE ($B1) is set to true; otherwise it is set to false. The BF (branch false) instruction transfers control to the instruction at location 10A (%F) if $B1 is false. (In the comments in Figure 14.13, instructions that are the targets of branches are labeled %.)

The remainder of the instructions is fairly straightforward, but to examine the addressing process in more detail, we examine the MOVE instruction above the ADD instruction labeled %C. The first operand address is 4F9E; we wish to determine how the actual storage is located. 9E represents a cell to be used as a subscript, and 4F represents something that is to be subscripted. The cell at 4F is a relocatable pointer that is the third substructure element relative to the cell at location 29. Looking at the cell at location 29, we find that it is a relocatable array that is the second substructure element relative to the

A TWO-MODULE EXAMPLE   221

```
Indexes                    Comments (assembly code)

130  01   090355846C   ACT   3,ESTAB,UNRESNAME,MATCHCODE
          16C002       MOVE  MATCHCODE,'2'
          004AE55      DEF   $B1,ESTAB
          FAE10A       BF    $B1,%F
          7AE19D5      EQ    $B1,TAG,'ESTB'
          9B022000     GT    $B2,SIZE,'0'
          AB222E4      LE    $B3,SIZE,'2000'
          05AEB0       AND   $B1,$B2
          05AEB2       AND   $B1,$B3
          FAE10A       BF    $B1,%F
          16C000       MOVE  MATCHCODE,'0'
          19E001       MOVE  I,'1'
     52   AAE9E22  %A: LE    $B1,I,SIZE
          7B06C000     EQ    $B2,MATCHCODE,'0'
          05AEB0       AND   $B1,$B2
          FAE10A       BF    $B1,%F
          004AE4F9E    DEF   $B1,BODY.ADDRESS(I)
          FAE0E1       BF    $B1,%C
          16C001       MOVE  MATCHCODE,'1'
          1A6001       MOVE  J,'1'
     88   AAEA622  %B: LE    $B1,J,SIZE
          7B06C001     EQ    $B2,MATCHCODE,'1'
          05AEB0       AND   $B1,$B2
          FAE0EB       BF    $B1,%D
          7AE3D9E3DA6  EQ    $B1,BODY.NAME(I),BODY.NAME(J)
          7B046A6B4    EQ    $B2,BODY.TYPE(J),MODULE
          7B046A6BF    EQ    $B3,BODY.TYPE(J),ENTRYPT
          06B0B2       OR    $B2,$B3
          05AEB0       AND   $B1,$B2
          FAE0E1       BF    $B1,%C
          16C000       MOVE  MATCHCODE,'0'
          14F9E4FA6    MOVE  BODY.ADDRESS(I),BODY.ADDRESS(J)
     E1   3A6001   %C: ADD   J,'1'
          E088         B     %B
     EB   7AE6C001 %D: EQ    $B1,MATCHCODE,'1'
          FAE100       BF    $B1,%E
          1843D9E      MOVE  UNRESNAME,BODY.NAME(I)
    100   39E001   %E: ADD   I,'1'
          E052         B     %A
239 10A   0A       %F: RETURN
```

Figure 14.13   Compiled instruction space for MATCHES.

cell at location 01. The cell at location 01 is a relocatable structure that is relative to the cell at location 55. The cell at location 55 is a parameter; this cell points to the argument, which is a pointer in this case. Hence, by following this chain of addresses and performing the necessary address calculations, the machine resolves the reference to the appropriate array element in the calling module.

Finally, note that the array in MATCHES is defined as containing 2000 elements, but the corresponding array in TESTEST has only eight elements. When an array element is being referenced and the array is passed as an argument (or, in this case, when the array is passed indirectly by passing a pointer variable on which a structure containing the array is based), the machine uses the defined bounds of the actual array to perform subscript checks. Hence a reference to BODY(9).NAME in procedure MATCHES would generate a bounds-exceeded fault.

SIGNIFICANCE OF SWARD

The SWARD architecture has satisfied its objectives of preventing or detecting significant sets of semantic and logic errors and providing enhanced error isolation. In addition, the branch and call traces and the fault-handling mechanism provide a base on which software testing and debugging tools can be built. In light of the increasingly critical nature of software reliability, these attributes alone make the architecture a significant advance.

To quantify some of these benefits, historical sets of software-error data were studied to measure the effectiveness of the architecture [1]. In one set of 91 PL/I, COBOL, and FORTRAN errors, 53 would be detected by the SWARD machine. Of the remaining 38 errors, 23 were associated with I/O functions in the languages. In another set of 39 errors, 17 would have been prevented or detected by the machine. Of a set of five common errors in PL/I list-processing applications, four would be detected by the SWARD machine, where only one of the five was detected by PL/I on the IBM S/370. Of a set of 38 errors monitored in the production of a PL/I application program, 18 would have been detected by the SWARD machine.

The benefits of such a machine are difficult, if not impossible, to quantify because the benefits of higher reliability are mostly intangible. However, an effort was made to quantify the benefits of the machine to the software development process. Using data on the relative frequencies of all types of programming errors and data on the probability distributions of when different types of errors are usually detected, it was estimated that a machine with this architecture could reduce a programming organization's testing and maintenance costs by 21% [1].

The architecture incorporates the concepts in Chapter 4 and thus significantly reduces the typically large semantic gap between programming languages and machine architectures. In fact, if Table 14.1 approximates typical circumstances, the SWARD machine appears to reduce the semantic gap to a greater extent than the earlier case studies. In addition to the concepts in Chapter 4, notable aspects of the architecture are the representation of sophisticated data structures, the ideas of supplemental instruction sets and cell types, the fault-handling mechanism, and variable-size addresses (the Burroughs B1700 also provides the latter).

The SWARD machine is currently an experimental architecture existing only on paper, and it is not without problems. One notable problem area is language imcompatibilities. For instance, FORTRAN, COBOL, and PL/I programs that make use of some existing machine-dependent aspect of the language cannot be easily compiled to this architecture. The most obvious incompatibility is the language construct that aliases variables with different attributes to the same storage area (e.g., use of the FORTRAN EQUIVALENCE statement, the COBOL REDEFINES clause, and the PL/I DEFINED attribute). There is no known way to represent such a construct in this architecture. On the other hand, there are many who consider the use of this construct to be poor programming style. Also, the architecture appears to be missing an important set of functions: input/output. However, this is not necessarily true; Chapter 16 discusses how input/output concepts might be added to this architecture using the existing instruction set and data types.

REFERENCE

1. G. J. Myers, "The Design of Computer Architectures to Enhance Software Reliability," Ph.D. dissertation, Polytechnic Institute of New York, 1977.

EXERCISES

14.1 What are the attributes and value of the cell D410237493

14.2 What is a major purpose of the token-string cell?

14.3 Given the following series of storage tokens, determine how many cells are represented and the attributes and values of each cell.

4003002010F31043729000A000A000000

14.4 Assume that the second MOVE instruction in Figure 14.11 is to be changed to assign 'A' to all elements of B.N. How would the machine instruction be represented?

14.5 How would the second MOVE instruction in Figure 14.11 be represented if the module's cell-address-size (CAS) field had the value 3?

14.6 Assume that the second MOVE instruction in Figure 14.11 is to be changed to assign 'A' to B (SIZE).N. How would the machine instruction be represented?

14.7 How would global data be represented in the architecture (e.g., a PL/I variable with the EXTERNAL attribute, a FORTRAN COMMON area)?

14.8 When would a compiler generate an UNDEFINE instruction?

14.9 Compile (on paper) the following FORTRAN program to the SWARD architecture.

```
      SUBROUTINE SORT (X,N)
      REAL X(N),SAVE
      INTEGER N,I,J
      IF (N .LT. 2) RETURN
      DO 20 I=2,N
        DO 10 J=1,I
          IF (X(I) .GE. X(J)) GO TO 10
          SAVE = X (I)
          X (I) = X (J)
          X(J) = SAVE
   10   CONTINUE
   20 CONTINUE
      RETURN
      END
```

You may wish to review the semantics of the FORTRAN DO statement first.

14.10 Consider a few instructions and possibly cell types that might be defined for a supplemental instruction set for COBOL, FORTRAN, or PL/I.

15 | SWARD Instruction Specifications

This chapter defines the basic instruction set of the machine. General notes that are applicable to many of the instructions are

1. Where an instruction permits two operands to be arrays, the arrays must be *conformable*. That is, they must have the same number of dimensions and the same number of elements in each dimension.
2. Where an instruction permits two operands to be structures, the structures must be conformable. That is, each structure must contain the same number of substructures. Corresponding elements in each structure must be compatible (as defined for the instruction).
3. Where an instruction specifies a particular cell type as a valid operand, the operand can also be a nested cell, unless otherwise noted. For instance, if an operand should be a decimal integer, the operand address can point to a decimal interger cell, an element in an array of decimal integers, a relocatable decimal integer, a decimal integer parameter, an element in an array parameter of decimal integers, and so on.
4. Most instructions can generate a common set of faults. For brevity, the set of fault types named the *general set* is defined as including the following faults: addressing, unknown data format, protection, invalid pointer, bounds-exceeded, invalid operand type, undefined operand, and incompatible operands.
5. "Arithmetic operands" are defined as the set—integer, literal,

fixed-point, and floating-point. "String operands" are defined as the set—boolean string, character string, and token string.
6. In the specifications of instruction formats, the first field is the operation code, which consists of 1–4 tokens, depending on the instruction. The abbreviation OA designates an operand address; IA designates an instruction address.
7. Literals are permitted as operand addresses, except where the instruction alters the operand's value or where the operand cannot be a decimal integer.

GENERAL INSTRUCTIONS

Instruction: MOVE

Function: The value of the second operand is moved into the first operand.

Format: 1,OA,OA

Operands: Both operands must be compatible, that is, both must be arithmetic, character or character string, boolean or boolean string, token or token string, or pointers. Both operands can be arrays, implying that an element-by-element move is done, or the first operand can be an array and the second not, meaning that the value of the second operand is moved into each element. Another valid case is when both operands are conformable structures and the corresponding elements are compatible. If the operands are arithmetic but have different types or sizes, the result is first converted to agree with the first operand. Rounding never occurs in the MOVE instruction. When a string is moved, the length of the first operand (if it is a string) is set equal to the length of the second operand. The operand combinations token string/character string and token string/character are compatible, and the combination character string/token string is compatible if the length of the token string is even.

Faults: General set (excluding invalid operand type) plus overflow and address loop.

Instruction: CONVERT

Function: The value of the second operand is moved into the first operand. A limited number of conversions may be done if the types of the two operands differ.

Format: 2,OA,OA

Operands: The rules of the MOVE instruction apply, but the rules concerning operand compatibility are somewhat relaxed. Table 15.1 de-

Table 15.1 Conversion Rules

		Operand2 type								
		di	dfx	dfl	b	c	t	bst	cst	tst
Operand1 type	di	1	1	1	2	3	2	2	4	2
	dfx	1	1	1					5	
	dfl	1	1	1					6	
	b	7			1	8	9	1	8	9
	c	10			11	1	12	11	1	12
	t	7			13	14	1	13	14	1
	bst	7			1	8	9	1	8	9
	cst	15	16	17	11	1	12	11	1	12
	tst	7			13	14	1	13	14	1

1 Acts identical to a MOVE instruction.
2 Converts it from binary to a positive integer.
3 The characters must be numeric (0–9).
4 All characters must be numeric except for the first, which can optionally be a + or −.
5 The string must be numeric optionally preceded by a + or −, or an optional + or − folowed by zero or more numerics followed by a . followed by zero or more numerics.
6 The string must be (1) numeric optionally preceded by a + or −, or (2) an optional + or − followed by zero or more numerics followed by a . followed by zero or more numerics, or (3) a number of form 2 followed by E, followed by an optional + or −, followed by one or two numerics.
7 Integer must be positive.
8 Character(s) must be 0 or 1.
9 Token(s) must be 0000 or 0001.
10 Produces the character(s) 0 to 9.
11 Produces the character(s) 0 or 1.
12 Produces the character(s) 0-9 and A-F.
13 Produces the token(s) 0000 or 0001.
14 Character(s) must be 0-9 and A-F.
15 Produces a string of numerics, preceded by a − if the number is negative.
16 Produces a string of the form numerics.numerics, preceded by a − if the number is negative.
17 Produces a string of the form 0.numericsEnumerics. If the number is negative, a − precedes the string. If the exponent is negative, a − follows the E.

scribes the valid conversions. A blank in the matrix indicates that no conversion will be performed and the incompatible-operands fault will occur. If a conversion is attempted but the value of the second operand does not meet the conversion rules, a conversion fault will occur.

Faults: General set plus conversion, overflow, and address loop.

Instruction: UNDEFINE (UNDEF)

Function: The value of the operand is set to undefined.

Format: 001, OA

Operands: The operand can be of any type. If it is a primitive token, the instruction has no effect. If the operand specifies a collection of cells (array or structure), each element receives the undefined value. If the operand is a string, its length is set to zero.

Faults: General set (excluding incompatible operands) plus address loop.

ARITHMETIC INSTRUCTIONS

Instruction: ADD

Function: The values of the two operands are added, and the result is placed in the first operand.

Format: 3,OA,OA

Operands: Both operands must be arithmetic. Both operands can be arrays, implying that an element-by-element addition is performed. If the first operand is an array and the second operand is a single number, the second operand is added to each element of the first operand. If the operands have different types or sizes, the value of the second operand is temporarily converted or adjusted to agree with the first operand before the addition is performed. The result is always rounded if least-significant digits will be lost. Floating-point results are always normalized.

Faults: General set plus overflow and underflow.

Instruction: SUBTRACT (SUB)

Function: The value of the second operand is subtracted from the value of the first operand, and the result is placed in the first operand.

Format: 4,OA,OA

Operands: See ADD instruction.

Faults: General set plus overflow and underflow.

ARITHMETIC INSTRUCTIONS

Instruction: MULTIPLY (MULT)

Function: The values of the two operands are multiplied and the result is placed in the first operand.

Format: 5,OA,OA

Operands: See ADD instruction.

Faults: General set plus overflow and underflow.

Notes: In the case of array operands, an element-by-element multiplication is done, not a "matrix multiplication."

Instruction: DIVIDE

Function: The value of the first operand is divided by the value of the second operand and the result (quotient) is placed in the first operand.

Format: 6,OA,OA

Operands: See ADD instruction.

Faults: General set plus overflow, underflow, and divide.

Instruction: MODULO

Function: The value of the first operand is divided by the value of the second operand, and the remainder is placed in the first operand.

Format: 002,OA,OA

Operands: Both operands must be decimal integers. Both can be arrays or scalars, or the first can be an array and the second a scalar.

Faults: General set plus overflow and divide.

Instruction: ABSOLUTE (ABS)

Function: The sign of the operand is set to positive.

Format: 003,OA

Operands: The operand must be arithmetic. If the operand is an array, the operation is performed on each element.

Faults: General set (excluding incompatible operands).

Instruction: COMPLEMENT (COMP)

Function: The sign of the operand is reversed.

Format: 01,OA

Operands: The operand must be arithmetic. If the operand is an array, the operation is performed on each element.

Faults: General set (excluding incompatible operands).

Instruction: POWER

Function: The value of the first operand is raised to the power given by the value of the second operand, and the result is placed in the first operand.

Format: 02,OA,OA

Operands: Both operands must be arithmetic. If the first operand is a decimal integer, the second operand must be a decimal integer. The first operand can be an array, implying that the operation is performed on each element. The result is always rounded if least-significant digits will be lost. Floating-point results are always normalized.

Faults: General set plus overflow and underflow.

COMPARISON INSTRUCTIONS

Instruction: EQUAL (EQ)

Function: If the values of the second and third operands are equal, the first operand (type boolean) is set to 1; otherwise it is set to zero.

Format: 7,OA,OA,OA

Operands: The first operand must be boolean. The second and third operands must be compatible (both arithmetic, character, boolean, pointer, or token). If the second and third operands are arithmetic but have different types or sizes, the value of the third operand is temporarily converted to agree with the second operand before the comparison is made. (Overflow faults never occur. If an overflow condition is encountered, the two operands are defined as unequal.) If the operands are strings of unequal length, the shorter string is temporarily padded with blanks (for character strings) or zeros (for boolean or token strings) before the comparison is made. Character and boolean strings are padded on the right, and token strings are padded on the left. The second operand may be an array, or the second and third operands may be arrays, in which case an element-by-element comparison is done. The result is true only if the relation holds between all corresponding elements. If the operands are pointers, only the logical addresses (not the access codes) are compared.

Faults: General set plus address loop.

Instruction: NOT-EQUAL (NE)

Function: If the values of the second and third operands are unequal, the first operand (type boolean) is set to 1; otherwise it is set to zero.

Format: 03,OA,OA,OA

COMPARISON INSTRUCTIONS

Operands: See EQUAL instruction.
Faults: General set plus address loop.

Instruction: LESS-THAN(LT)
Function: If the value of the second operand is less than the value of the third operand, the first operand (type boolean) is set to 1; otherwise it is set to zero.
Format: 8,OA,OA,OA
Operands: The first operand must be boolean. The second and third operands must both be arithmetic, character, or token. If they are arithmetic but have different types or sizes, the value of the third operand is temporarily converted to agree with the second operand before the comparison occurs. (Overflow faults never occur. If an overflow condition is encountered, the second operand is taken as being less than the third.) If the second operand is a string, the third operand must be a string of equal type. Character strings are compared based on the collating sequence of characters (EBCDIC representation). Token strings are compared by viewing them as positive hexadecimal numbers. Unequal-length strings are padded as described in the EQUAL instruction. The second operand may be an array, or the second and third operands may be arrays, in which case an element-by-element comparison is done. The result is true only if the relation holds between all corresponding elements.
Faults: General set.

Instruction: GREATER-THAN (GT)
Function: If the value of the second operand is greater than the value of the third operand, the first operand (type boolean) is set to 1; otherwise it is set to zero.
Format: 9,OA,OA,OA
Operands: See LESS-THAN instruction.
Faults: General set.

Instruction: LESS-THAN-OR-EQUAL (LE)
Function: If the value of the second operand is less than or equal to the value of the third operand, the first operand (type boolean) is set to 1; otherwise it is set to zero.
Format: A,OA,OA,OA
Operands: See LESS-THAN instruction.
Faults: General set.

Instruction: GREATER-THAN-OR-EQUAL (GE)

Function: If the value of the second operand is greater than or equal to the value of the third operand, the first operand (type boolean) is set to 1; otherwise it is set to zero.

Format: 04,OA,OA,OA

Operands: See LESS-THAN instruction.

Faults: General set.

Instruction: DEFINED (DEF)

Function: If the value of the second operand is defined, the first operand (type boolean) is set to 1; otherwise it is set to zero.

Format: 004,OA,OA

Operands: The second operand can be of any type. If the operand is a primitive token, the result is always 1. If the second operand specifies a collection of data (array or structure), the first operand is set to 1 only if every element has a defined value. If the second operand is a string and its length is zero, the first operand is set to zero.

Faults: General set (excluding incompatible operands). The undefined-operand fault will not occur unless the operand is located indirectly (i.e., through a parameter or pointer) and the parameter or pointer cell has the undefined value.

BOOLEAN INSTRUCTIONS

Instruction: AND

Function: The values of the two operands are "anded," and the result is placed in the first operand.

Format: 05,OA,OA

Operands: The operands must both be boolean, equal-length boolean strings, or arrays of booleans or boolean strings. The first operand may be an array, or the first and second operands may be arrays.

Faults: General set.

Instruction: OR

Function: The values of the two operands are "or-ed," and the result is placed in the first operand.

Format: 06,OA,OA

Operands: See AND instruction.

Faults: General set.

STRING INSTRUCTIONS

Instruction: NOT

Function: The value of the boolean operand is inverted.

Format: 005,OA

Operands: The operand must be boolean, a boolean string, or an array of booleans or boolean strings.

Faults: General set (excluding incompatible operands).

STRING INSTRUCTIONS

Instruction: CONCATENATE (CONCAT)

Function: The second operand is concatenated to the first operand.

Format: B,OA,OA

Operands: The two operands must be strings of the same type. The length of the first operand is incremented by the length of the second operand, and the value of the second operand is appended to the end of the first operand.

Faults: General set plus overflow.

Instruction: MOVE-SUBSTRING (MOVESS)

Function: The substring (part of a string) designated by the second set of operands is moved into the substring designated by the first set of operands.

Format: C,OA,OA,OA,OA,OA

Operands: Operands 2 and 4 designate the two strings. The two operands must be compatible (both character, boolean, or token). Operands 1, 3, and 5 must be decimal integers. Operand 1 specifies the length of the substring to be moved. Operand 3 specifies the index of the start of the substring in the target string, and operand 5 specifies the index of the start of the substring in the source string.

Faults: General set.

Notes: MOVESS performs an overlay rather than an insertion. That is, the length of the target string is unchanged.

Instruction: INDEX

Function: A string is searched for a specified substring. If the substring is found, the first operand contains the index of the start of the matching substring in the string. If the substring is not found, the first operand is set to zero.

Format: 07,OA,OA,OA

Operands: The first operand must be decimal integer. The second operand is the string to be searched. The third operand must be a string having the same type as the second operand; it represents the substring to be located.

Faults: General set plus overflow.

Instruction: LENGTH

Function: The length of the second operand (a string) is placed in the first operand.

Format: 08,OA,OA

Operands: The first operand must be a decimal integer, and the second operand must be a string.

Faults: General set (excluding incompatible operands and undefined operand) plus overflow.

CONTROL INSTRUCTIONS

Instruction: CALL

Function: Execution of the current module is suspended, and execution of another module begins at the specified entry point. Allocation and initialization of automatic storage is performed for the called module.

Format: D,OA,OA,X,Y,OA1, . . . ,OAX

Operands: The first operand is a pointer to the called module (must have read access). The second operand is a decimal integer specifying the number of the entry point in the called module. The first immediate field is a 2-token hexadecimal number (X) specifying the number of arguments to be passed. The second immediate field is a 2-token hexadecimal number (Y) (from 0 to X) specifying that the called module is restricted to read-only access to the first Y arguments. The last X operands are the arguments. Argument address fields cannot be literals.

Faults: General set (excluding incompatible operands) plus call-trace and invalid transfer.

Notes: The CALL instruction does not actually transfer arguments to the corresponding parameters in the called module. This must be done via an ACTIVATE instruction at the called entry point. CALL creates an activation record and places it on the top of the stack of the activation records for the program.

CONTROL INSTRUCTIONS

Instruction: ACTIVATE (ACT)

Function: Execution of a module is initiated, consisting of the initialization of parameters and checking of argument-parameter compatibility.

Format: 09,X,OA1, . . . ,OAX

Operands: The immediate field (X) is a 2-token hexadecimal number specifying the number of parameters. The X operands are the parameters (all must be of type parameter). Literal and array element operands are prohibited.

Faults: Addressing, unknown data format, invalid pointer, invalid operand type, invalid parameter count, incompatible operands, and address loop.

Notes: The rules for argument-parameter compatibility are defined in Table 15.2.

Instruction: RETURN

Function: Execution of the current module is terminated, and execution is resumed after the CALL instruction that called this module.

Table 15.2 Rules for Argument-parameter Compatibility

If the Argument Is	Then the Parameter Must Be
Integer, size=N	Integer, size $\geqslant$ N or fixed-point, size-fsize $\geqslant$ N or fixed-point, size $\geqslant$ N
Fixed-point, size=N, fsize=M	Fixed-point, size $\geqslant$ N, fsize $\geqslant$ M, size-fsize $\geqslant$ N-M or floating-point, size $\geqslant$ N
Floating-point, size=N	Floating-point, size $\geqslant$ N
Boolean	Boolean
Character	Character
Token	Token
Boolean string, size=N	Boolean string, size $\geqslant$ N
Character string, size=N	Character string, size $\geqslant$ N
Token string, size=N	Token string, size $\geqslant$ N
Pointer	Pointer
Structure	Structure (must be conformable and each element must be compatible)
Array	Array (with same number of dimensions; element type must be compatible)

Format: 0A

Operands: None.

Faults: None.

Notes: RETURN "undoes" the effect of the previous CALL and ACTIVATE instructions. That is, the current activation record is destroyed.

Instruction: LOCAL-CALL (LCALL)

Function: Execution is suspended, and control is transferred to an instruction within the module.

Format: 0B,IA,X,Y,OA1, . . . ,OAX

Operands: The first address field specifies an instruction address to which control is transferred. The remaining fields are identical to those of the CALL instruction.

Faults: Addressing, unknown data format, bounds-exceeded, invalid transfer, call-trace, and invalid pointer.

Notes: LCALL, unlike CALL, does not create an activation record, which means that internal procedures cannot be recursive (unless the compiler uses an ALLOCATE instruction to simulate the effect of an activation record), and that any scope-of-name rules are the compilers' responsibility.

Instruction: LOCAL-ACTIVATE (LACT)

Function: Execution of an internal procedure is initiated, including initialization of parameters and checking of argument-parameter compatibility.

Format: 0C,X,OA1, . . . ,OAX

Operands: The immediate field (X) is a 2-token hexadecimal number specifying the number of parameters. The X operands are the parameters (all must be of type parameter). Literal and array element operands are prohibited.

Faults: Addressing, unknown data format, invalid pointer, invalid operand type, invalid parameter count, incompatible operands, and address loop.

Notes: The rules for argument-parameter compatibility are the same as those for the ACTIVATE instruction.

Instruction: LOCAL-RETURN (LRETURN)

Function: Execution is transferred to the instruction following the last LCALL instruction executed in the current module.

ADDRESSING INSTRUCTIONS

Format: 0D

Operands: None.

Faults: Invalid transfer (if there was no previous LCALL instruction).

Notes: If LRETURN is executed in a fault handler and if there is no outstanding LCALL instruction, execution of the fault handler is terminated and execution resumes at the faulting instruction.

Instruction: BRANCH (B)

Function: Control is transferred to the designated instruction.

Format: E,IA

Faults: Invalid transfer and branch-trace.

Instruction: BRANCH-FALSE (BF)

Function: If the first operand (boolean) has the value zero, control is transferred to the specified instruction. Otherwise control is transferred to the instruction following the BF.

Format: F,OA,IA

Operands: The operand must be boolean.

Faults: General set (excluding incompatible operands) plus invalid transfer and branch-trace.

ADDRESSING INSTRUCTIONS

Instruction: CREATE-POINTER (CPTR)

Function: The first operand is assigned the logical address of the second operand.

Format: 0E,OA,OA

Operands: The first operand must be a pointer. The second operand may be any operand. The access code in the pointer is set to the access that the module currently has to the second operand. The logical address is the address of the value of the second operand. If the second operand is a relocatable or parameter, the address of the storage addressed by these cells is computed and assigned to the first operand.

Faults: Addressing, unknown data format, invalid pointer, bounds-exceeded, invalid operand type, and protection.

Instruction: CHANGE-ACCESS (CACC)

Function: The access code in the first operand is reduced to the value specified.

Format: 006,OA,X

Operands: The first operand must be a pointer. The immediate field (X) is 1 token in length. The access code in the pointer is changed to the lesser of its current value or the value of the immediate field.

Faults: General set (excluding incompatible operands).

Instruction: ALLOCATE (ALLOC)

Function: An area of storage is allocated for the operand, and the pointer associated with the operand is initialized.

Format: 0F,OA

Operands: The operand must be relocatable and can describe any type of cell. The cell-based-upon field in the operand's tag must point to a pointer cell, and the substructure-number field in the tag must be zero. All allocated storage is initialized to the value undefined. The access code in the pointer is set to 1110 (read/write/free access). If the storage is not explicitly freed, it is automatically freed when the program ends.

Faults: Addressing, unknown data format, protection, and invalid operand type.

Instruction: FREE

Function: The storage object specified by the operand is freed.

Format: 007,OA

Operands: The operand must be a relocatable or pointer cell. If it is relocatable, the cell-based-upon field in its tag must point to a pointer cell and the substructure-number field in the tag must be zero. The storage referenced by the pointer is freed. If the operand is a pointer, the storage referenced by the pointer is freed. In both cases the pointer must have "free" access. The pointer is given the undefined value at the end of the instruction.

Faults: General set (excluding incompatible operands).

Notes: The pointer must name an entire storage object. For example, attempting to free a cell within an activation record or a cell *within* a dynamically allocated storage area would result in a protection fault.

Instruction: CHAIN

Function: The operand is a pointer that may point to a chain of pointers. The operand is assigned the value of the last pointer in the chain.

Format: 008,OA

ADDRESSING INSTRUCTIONS

Operands: The operand must be a pointer. If the pointer references a cell other than a pointer, the instruction does nothing. If the pointer points to another pointer, the chain of pointers is followed. The operand is assigned the value of the last pointer (i.e., the first pointer encountered that points to a nonpointer cell). The operand is assigned the *least* (most restrictive) access code of all the pointers in the chain.

Faults: General set (excluding incompatible operands) plus address loop.

Instruction: LOAD-MODULE (LMODULE)

Function: The specified module is loaded (defined to the machine), and a pointer to the module is assigned to the other operand.

Format: 009,OA,OA

Operands: The first operand must be a pointer, and the second operand must be a token string. The token string must have the form of a module (see Figure 14.6). The machine checks the validity of the format of the module, copies it into internal storage, and creates a pointer to it. The pointer is assigned read/write/free access (but writing into a module, except with the LINK instruction, is prohibited by the machine).

Faults: General set (excluding incompatible operands) and invalid module.

Notes: The validity checking that is done by this instruction is not specified here.

Instruction: LINK

Function: A pointer value is assigned to a specified pointer cell in a loaded module.

Format: 00A,OA,OA,OA

Operands: The first operand is a pointer to a loaded module. The pointer must have write access. The second operand is a 5-token string that specifies a cell address in the loaded module. The third operand is a pointer. The value of this pointer is assigned to the pointer cell specified by the first and second operands. The cell must be in the static or constant storage die.

Faults: General set (excluding incompatible operands). The addressing fault will occur if the target cell is not in the static or constant storage die.

DEBUGGING INSTRUCTIONS

Instruction: ENABLE

Function: The specified token string is "or-ed" into the fault-code field as defined in the module header. The fault-code field is maintained in the module's activation record.

Format: 00B,OA

Operands: The operand must be a token string (of length A) whose length is equal to or less than the length of the fault-code field. If the token string is shorter than the fault-code field, only the first A tokens of the fault-code field are changed.

Faults: General set (excluding incompatible operands) and overflow.

Instruction: DISABLE

Function: The inverse (negation) of the specified token string is "and-ed" into the module's fault-code field in the activation record.

Format: 00C, OA

Operands: See ENABLE instruction.

Faults: General set (excluding incompatible operands) and overflow.

Instruction: SIGNAL

Function: A signal fault occurs. The 2-token immediate field is transmitted to the fault handler.

Format: 00D,X

Faults: Signal.

Instruction: CONTINUE (CONT)

Function: Execution of the fault handler is terminated, and execution resumes at the instruction following the faulting instruction.

Format: 00E

Faults: Invalid transfer (if there is no current fault, if continuing beyond the current fault is not permitted, or if CONTINUE is issued from a subroutine called by a fault handler).

Notes: If a fault handler wishes to resume execution at the faulting instruction, it should issue the LRETURN instruction. If the fault handler wishes to resume execution at the instruction following the faulting instruction, it should issue the CONTINUE instruction. The only faults that may be followed by a CONTINUE instruction are incompatible operands, overflow, underflow, divide, and conversion.

DEBUGGING INSTRUCTIONS

Instruction: TRANSFER-FAULT (TRFAULT)

Function: The current fault handler is terminated and a higher fault handler (one lower in the activation stack) is called. If an applicable fault handler cannot be found, the program is terminated.

Format: 00F

Faults: Invalid transfer (same first and third situations as in the CONTINUE instruction).

Notes: TRFAULT is used by a fault handler that has a particular fault enabled, but after receiving such a fault it decides to send it to a "higher authority."

Instruction: DISPLAY-TAG (DTAG)

Function: The tag of the designated cell is assigned to a token string operand.

Format: 0001,OA,OA,OA

Operands: The second operand is a pointer to a loaded module; the pointer must have read access. The third operand is a 5-token string that specifies a cell address in the loaded module. The tag of this cell is moved into the first operand, which must be a token string. The overflow fault is suppressed; if the tag is longer than the first operand, the first operand is filled with the leftmost tokens of the tag. If the pointer is undefined, it is assumed to designate the current module (i.e., an undefined-operand fault will not occur for the second operand).

Faults: General set (excluding incompatible operands).

Notes: This instruction is intended only for use by debugging functions. For planning purposes, the largest possible tag is 1278 tokens (a 15-dimension array of structures of 255 substructures where the cell address size is 5 tokens).

Instruction: DISPLAY-CONTENTS (DCON)

Function: The contents component of the designated cell is assigned to a token string operand.

Format: 0002,OA,OA,OA

Operands: See DTAG instruction. Overflow faults are similarly suppressed. If the cell is in the automatic storage die, its value for the most recent, currently active activation of the module is displayed. If the module is not active, the cell's initial value in the die in the module is displayed.

Faults: General set (excluding incompatible operands).

Notes: This instruction returns the *contents* of a cell, which is not always identical to its value. For example, the contents of an array cell is just a 6-token internal machine field; the contents of a character string is a 3-token length field and a variable-size value; the contents of a pointer is a 1-token access code and a 20-token logical address. The size of a contents component can be determined by first using a DTAG instruction.

Instruction: TRACE

Function: A specified trace is enabled for a specified module or modules.

Format: 0003,X,OA

Operands: The 1-token immediate field (X) specifies the type of trace. The value 0001 specifies a branch trace, 0010 specifies a call trace, and 0011 specifies both. The second operand must be a pointer or an array of pointers. The pointers must point to modules and must have write access. The trace is enabled in the modules specified by the pointers. If a pointer is undefined, it is assumed to designate the current module.

Faults: General set (excluding undefined operand and incompatible operands).

Notes: If a branch trace is specified, a branch-fault occurs whenever a B or BF instruction is executed in the designated modules. If a call trace is specified, a call-trace fault occurs whenever a CALL or LCALL is executed.

Instruction: NOTRACE

Function: A specified trace is disabled for a specified module or modules.

Format: 0004,X,OA

Operands: See TRACE instruction. If a trace was not previously enabled in a module, disabling it has no effect.

Faults: General set (excluding undefined opeand and incompatible operands).

CALCULATION OF THE ADDRESS-FIELD SIZE

The use of variable-size address fields places a burden on the compiler in the form of determining the appropriate size of the address field for the module being compiled. Of course a simple-minded compiler need

CALCULATION OF THE ADDRESS-FIELD SIZE

not face up to this burden; it could simply use a fixed-size address field that is large enough for the largest module that can be compiled, but such a solution does not exploit the advantages of variable-size addresses.

The address field size is a function of the size of the module's address space (the size of the four dies in the module), but this calculation is not straightforward because some cell types in the address space contain address fields (in particular the cell types *structure* and *relocatable*). The formula for calculating the smallest address field is

$$N = \text{CEIL}(\log(N(F+R)+3S+3R+8A+4D+22P+E+Z-Q-6L))$$

The function CEIL raises a number to the next higher integer. All logarithms are base 16. The identifiers have the meanings:

N—number of tokens in the address field.
S—number of structure tags (including nested tags).
R—number of relocatables.
A—number of arrays.
D—total number of array dimensions in all arrays.
P—number of parameters.
E—number of tokens of die space for all cells that are not arrays, parameters, relocatables, and structures.
Z—size of the variable-size subtag fields in tags of type array, parameter, and relocatable (but excluding subtags for arrays and structures).
F—number of substructure fields in all structure tags and subtags.
Q—size of last cell in last die.
L—number of relocatable arrays.

Since N appears on both sides of the equation and the equation cannot easily be solved for N, the best approach is to substitute the values 2, 3, 4, and so on for N until both sides are equal.

The other type of variable-size address is the instruction address. The formula for calculating the smallest instruction address needed is

$$M = \text{CEIL}(\log(B + MI))$$

where

B—number of tokens in the instruction space excluding all instruction address fields and excluding the last instruction.
I—number of instruction address fields.

Compiler Considerations

In producing a compiler for this architecture these approaches are available:

1. Use fixed large values for N and M. This is the simplest approach, but it does not take advantage of the use of shorter addresses.
2. Use the formulae for N and M to find the optimal sizes. This approach takes full advantage of the encoding, but it complicates the compilers.
3. Rather than using the formulae, use a few simple heuristics to guess at the optimal N and M. If, during code generation, the compiler finds that N or M is too small, increment it by 1 and begin the code generation again.
4. Choose constant values for N and M. For instance, N=4 seems to be a reasonable upper bound, for it defines an address space of a maximum of 65535 tokens (which seems even more reasonable considering the fact that space for array elements does not appear in the address space). A separate optimization or "module-compression" program can then be written that is compiler and language independent. Its function is to take a module with a possibly oversized address field and produce an equivalent module with a minimal address field.

INTERNAL STORAGE OBJECTS

This section is not part of the SWARD architecture because it describes the internal storage objects maintained by the machine. The content and format of these objects is an implementation, not architectural, consideration, because this information is transparent to programs coded to the architecture. However, the information is included here because it may assist in understanding the architecture. The internal storage objects shown here are taken from a software simulator of the architecture.

The activation record is shown in Figure 15.1. The activation record, created when a module is called, contains status information and the module's automatic storage. The activation record begins with an 8-bit code (ARID) that uniquely defines the storage area as an activation record.

Most fields in the activation record should be self-explanatory. The ninth field, an address of a chain of all dynamically allocated storage

Figure 15.1 SWARD internal storage objects.

areas, is used only in the first activation record of each program. The machine uses this field to free any allocated storage areas remaining when a program terminates. The format of the dynamically allocated storage object is also shown in Figure 15.1.

As mentioned in the architecture specification, the LCALL instruction does not create an activation record. Instead, it creates entries on an LCALL stack, which is pointed to by the activation record. The LCALL-stack entry, as shown in Figure 15.1, contains information related to fault handling. The reason is that fault handlers are entered as if they were internal subroutines. That is, an LCALL-stack entry is created when an LCALL instruction is executed or when a fault handler is entered. The LCALL-stack entry is not one of the storage objects mentioned in the architecture specification (i.e., it does not have a logical address), but it is shown here for clarification.

The last object in Figure 15.1 is an internal module. An internal-module object is created as the result of a LOAD-MODULE instruction. Note that the format of an internal module is almost identical to that of the architected module format shown in Figure 14.6, but this need not be the case. For instance, the LOAD-MODULE instruction could make transformations on the internal module, such as optimizing the instructions in the instruction space. Thus the actual instructions executed by the processor need not be the same instruction set as specified earlier in the chapter, providing that the machine maintains the illusion of executing the architected instruction set. However, no advantages were found in doing this in the simulator (the instruction set described earlier was considered optimal for direct interpretation); thus in this implementation the form of data and instructions in the internal module is identical to their form in the architected definition of the module.

As an aside, a subset of the error checking performed by the SWARD machine could be performed by the LOAD-MODULE instruction (i.e., when the internal module is being constructed), but currently it is not.

VI
Related Topics in Computer Architecture

16 | Input/Output Architecture

Given the major role of input/output in today's computer applications, it is alarming to note that the input/output architectures (i.e., the abstraction of input/output data, devices, and media that is presented to machine-language programs) in current systems do not differ significantly from earlier systems. In fact, even most of the non-von Neumann machines possess input/output architectures that bear a close resemblence to systems of the 1950s.

It is fair to say that most current systems have input/output architectures that are oriented toward the earliest physical input and output devices: sequential punched-card readers and punches. Newer input/output storage forms, such as magnetic tape and rotating magnetic disks, were retrofitted into the sequential punched-card interface rather than introduced through major architectural changes. After software file-management systems were constructed on this largely sequential interface, data base management systems evolved, but rather than leading to changes in the underlying architecture, they were built on the file-management systems.

This "add-on" evolution in both software and hardware input/output architectures has led to two problems in current computing systems. First, similar to the semantic gap between programming languages and computer architectures discussed in Chapter 2, there is a large semantic gap between a program's input/output requests and the primitive representation of input/output media at the hardware/software interface. As an example, consider the semantic gap between an application program's request to the underlying data base management system to "obtain the names of all employee's in division X1 whose salary is greater than their immediate manager's salary and whose time to retirement is less than 10 years" and the I/O operations in the underlying machine (e.g., channel command words in the IBM S/370) such as "move read/write heads to disk cylinder 47" and "read record 4 on track 7." This semantic gap is usually bridged by many layers of interpretive software (e.g., data base manager, file handler, device handler, I/O scheduler), leading to significant overhead. Furthermore, the processing of such an application-program request might involve the movement of literally millions of bits of information (e.g., index and data records) between the I/O devices and main memory, another tremendous source of overhead.

Second, current systems do not present a unified view of the concept of memory. If a datum resides in main storage, it is addressed in one fashion, but if the datum resides on a magnetic disk, it is addressed in a completely different fashion. Not only does this situation exist at the computer-architecture level (e.g., the use of machine instructions versus channel command words in the IBM S/370), but it exists at the programming-language level (e.g., the use of an assignment statement versus a GET statement in PL/I). Hence what could be thought of as a single concept—memory—is usually represented as a set of apparently unrelated concepts, leading to additional software and computer-architecture complexity.

To be fair, some systems have taken steps, although only partially, toward the solution of the second problem. Systems with virtual memories allow programs to be unaware of whether a datum is in main memory or on a back-up storage device, although, in most systems, this is primarily a software, rather than a computer architecture, mechanism, and it does not apply to all storage (e.g., files). Systems with high-speed cache memories (buffers between the processor and main storage) usually implement this as a transparent concept. Systems like the IBM 3850 give programs the illusion that all files are stored on magnetic disks, but whole files or pieces thereof may be staged to and from magnetic tapes. However, none of these steps completely unifies the architectural concept of memory.

FRONT- AND BACK-END PROCESSORS

It has long been recognized that systems containing large data bases and/or large numbers of communication lines expend a significant

FRONT- AND BACK-END PROCESSORS

amount of resources (e.g., central processor time) managing these other resources. An obvious deduction, then, is that one might substantially improve system performance by distributing this function across multiple processors, that is, by partitioning the communications functions to specialized "front-end" processors and partitioning the data and memory management functions to specialized "back-end" processors.

The motivation for a back-end data-management processor is to remove all or most of the data and memory management functions from the central processor, thus reducing its load [1–4]. For instance, the Control Data Star-100 system can have one or more "file-storage stations," which communicate with the central processor via messages [4]. The file-storage stations perform logical operations on files when directed to do so by the central processor (e.g., opening and closing files and accessing records) as well as physical operations (e.g., memory management) and auxiliary functions (e.g., security, recovery, and the collection of performance and accounting statistics). The basic advantages of such back-end processors are that they facilitate the sharing of a common data base among multiple systems or processors (by providing a common control mechanism through which all data operations are routed), they provide additional data security (by establishing another "filter" that resides in a separate machine), and, as mentioned previously, they enhance system performance by reducing the instruction-processing load of the central machine and the amount of data transferred into main storage.

The motivation for front-end processors is similar. Functions such as the handling of communication protocols, message compression and decompression, polling of terminals, and the management of communication errors are placed in one or more front-end processors, thus reducing the load on the central processor. The advantages of front-end processors have been recognized for some time, making them fairly commonplace on many current systems.

Although this distribution of function to front- and back-end processors may be worthwhile, it is not pursued further in this book because it is usually not related to computer architecture. The placement of a data base management system into a back-end processor and a telecommunications access method into a front-end processor are issues of configuration architecture (as defined in Chapter 1) rather than computer architecture. In the way the partitioning is usually performed, it has no effect on the system's hardware/software boundary. Furthermore, these configuration issues are not of interest here because they have no positive effect on the two problem areas noted earlier—the input/output semantic gap and the unified concept of memory.

ASSOCIATIVE-STORAGE PROCESSORS

One fundamental consideration in the method by which storage is addressed is the realization that addressing by location, the traditional method, is often inappropriate. For instance, addressing by location is appropriate for the statement "add variable B to variable A," but it appears to be of little utility for the statement "obtain all employees earning more than $20,000 per year."

Unfortunately, conventional systems present only a sequential view of storage and only permit storage to be addressed by location (e.g., read record 12 on track 4 on cylinder 73 on disk 5). Systems that process requests such as the second request are led to the need for large, complex, and inherently slow programs (data base managers) to transform tree, network, and other logical views of data into sequential storage media. Since requests like the one above prove to be intolerably slow if done via a sequential search, these programs have required the creation of additional storage structures called access paths (e.g., indexes). An index (providing that the need for the particular index has been anticipated in advance) can appreciably reduce the overhead of a search, but on the other hand it introduces more overhead when additional data items are added. It also increases the storage space needed, and it increases further the complexity of these programs.

One solution to this dilemma is the realization of the need to address storage by *content*, or associatively. That is, rather than saying "tell me what is stored in location X," one might wish to reverse the mechanism by saying "tell me the storage locations in which field Y (e.g., bits 7–35) has a value greater than 20,000." A storage with this property is called a *content-addressable storage* or an *associative storage*.

An associatively addressed memory has the properties: (1) accesses to storage cells (e.g., words or records) occur in parallel (i.e., memory operations refer to the collection of cells rather than a single cell), (2) memory operations are simultaneously applied to each storage cell, (3) searching or comparison is a primitive memory operation, and (4) search time is independent of the number of cells in the memory.

A simplified model of an associative storage device is illustrated in Figure 16.1. The storage array contains the cells (e.g., data base records). The comparand register contains the current search argument (e.g., the value 20,000 if one is about to locate the employees earning $20,000 or more). The mask register indicates the bit substrings or fields of the storage cells to which the search argument is to be applied. The comparand-operation register contains the current search operation; typical operations might be equal, not equal, maximum, minimum,

ASSOCIATIVE-STORAGE PROCESSORS

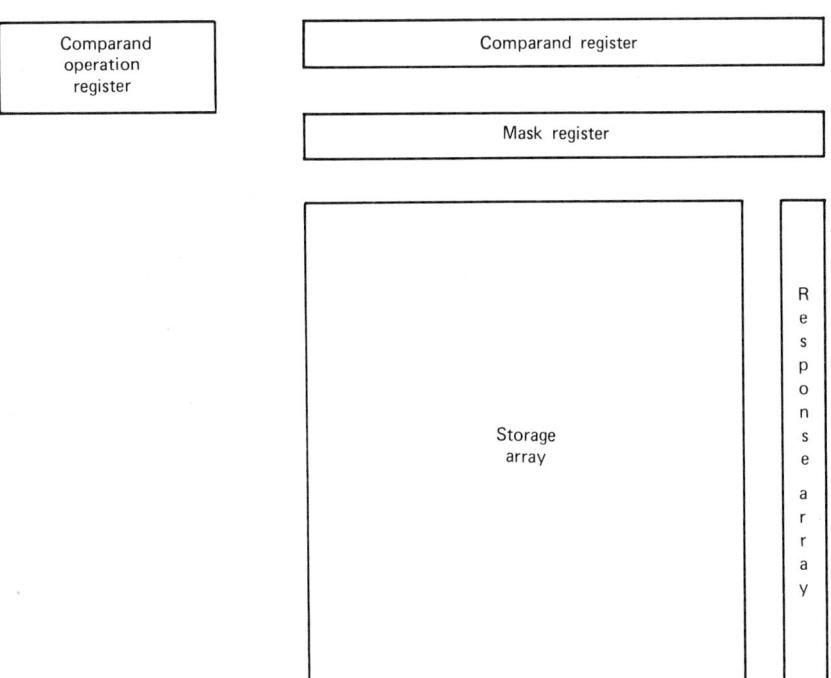

Figure 16.1 Model of an associative store.

between, next higher, next lower, less than, and so on. The response array contains 1 bit per storage cell; it is used to mark all cells meeting the search criteria. As is indicated in Figure 16.1, the comparand and mask registers extend over the response array, implying that search operations can be applied to the response array as well as the storage array. That is, the set operations of intersection, union, and complement can occur among successive searches by allowing operations to be dependent on the prior state of the response array.

To illustrate the use of such a storage device, assume that each storage cell represents an employee record and that bits 7–26 contain the annual salary. To accomplish the task of finding all employees making greater than $20,000 and less than or equal to $25,000, the instructions to the device might be

1. SEARCH MASK=7-26 GREATER THAN 20000
2. SEARCH (MASK=7-26 LESS THAN OR EQUAL 25000) AND (MASK=RA EQUAL TRUE)
3. READ MASK=RA EQUAL TRUE

In processing instruction 1, the device will set the response-array (RA) bit for all corresponding cells whose salary field (bits 7–26) is greater than 20,000; the response-array bits for all other cells are reset (to zero). Each cell is tested simultaneously, implying that the time of the instruction is independent of the number of cells. In processing instruction 2, each cell's response bit is set providing that the salary field is 25,000 or less and the response bit is already set. All other response bits are reset to zero. Instruction 3 fetches all cells whose response bit is set (i.e., all cells whose salary field is greater than 20,000 and less than or equal to 25,000).

Essentially, then, an associative memory appears to have a substantial amount of logic (or a miniature processor, so to speak) attached to each storage cell. One consequence of this is the impression that associative memories must be extremely costly, but this impression is not necessarily true for two reasons. First, the economics of circuit technology are changing such that the design cost of a circuit is much more significant than the cost of producing it, providing that it is manufactured in large quantities. The advantage of an associative memory is that the system is iterative. The design cost of the associative logic is amortized over each of the cells (perhaps thousands), thus reducing the cost of each cell. Second, tradeoffs are possible such that the associative logic is not tied to individual storage cells, but perhaps to small groups of cells. For instance, the memory could be constructed as a set of shift registers where each shift register contains a set of storage cells and one set of associative logic.

In the same vein, the most promising use of associative addressing appears to be for rotating memories (e.g., magnetic disks), where each read/write head contains associative logic. Rather than examining each record in parallel, a record is examined as it passes a read/write head. To see the utility of this, assume that a large data base is spread over 100 fixed-head disks with a speed of 100 revolutions per second. Assume that each disk contains 1000 tracks, that each track contains 100 records, and that an associative operation is being applied to three fields in each record (e.g., search for all taxpayers who were born in 1946, earn less than $15,000, and are unmarried.) This associative operation has an execution time of 10 milliseconds (one rotation of the memory). Its comparison rate is 3 billion per second, far exceeding the speed of any sequential processor. In addition, this associative storage has the following advantages over a conventional multi-indexed data base:

1. Far less storage space is needed (no storage is needed for indexes).
2. Significantly less data must pass between the storage and the processor (significant by many orders of magnitude).

3. The load on the central processor is significantly reduced.
4. The retrieval time is constant (i.e., independent of the number of records).
5. In addition to retrieval being fast, updates (e.g., adding new records) are fast. A record is added by simply writing it on a track; no indexes must be updated.
6. The software data-management program is considerably simplified.
7. The system is easier to expand. If the data base grows, the same performance can be attained by simply adding more associative disks rather than requiring the addition of a faster (and possibly unattainable) central processor.

The number of experimental associative memories is too large to mention all of them, but as representative examples, the STARAN system is a commercial machine containing up to 32 associative arrays, each array containing 256 256-bit words [5]. ECAM is a system containing an associative memory of up to 250,000 shift registers, each 4096 bits in length [6]. A back-end machine employing associative memories and emphasizing data security has been designed [7]. The structure of several proposed associative, rotating-disk memories have been described [8–10]. Another system, RAP (Relational Associative Processor), is based on associative rotating disks where many of the fundamental operations of the relational data base concept [11,12] are primitive system instructions [13]. This system is discussed in more detail in the next section. Finally, DeFiore and Berra present a quantitative analysis of conventional inverted-list memories versus associative memories [14], and Langdon discusses some engineering considerations of associative, rotating memories [15].

THE RELATIONAL ASSOCIATIVE PROCESSOR

To illustrate in more detail the architecture of an associative memory, the RAP (Relational Associative Processor) is used as an example [13,16–19]. RAP is an associative memory subsystem based on a rotating bulk memory and oriented toward the relational view of a data base [11,12].

Figure 16.2 is a picture of the overall structure of RAP. The general-purpose computer communicates to RAP by sending it a RAP program. A RAP program is a sequence of RAP instructions expressing a simple or complex data base request. The controller is a processor that decodes RAP programs, controls the associative logic, and communicates with the general-purpose computer. The set processor is a specialized processor that participates in set operations involving the entire data base.

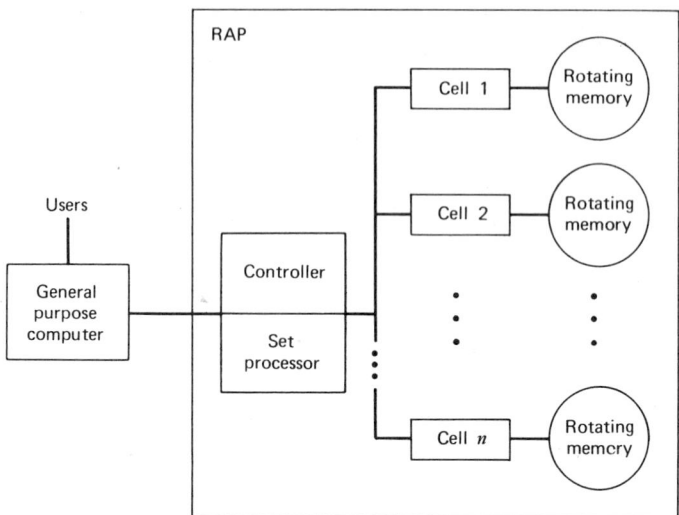

Figure 16.2   Structure of RAP.

The associative characteristic is provided by the units named cells. A cell consists of a microprocessor and an associated rotating memory. The rotating memory can be a track of a fixed-head disk or drum, or a circular shift register, but for the discussion here it is assumed to be a track of a rotating disk. The RAP unit might be expected to contain hundreds or thousands of cells. The cells, when directed by the controller, perform search, manipulation, and numerical operations on their memories.

When used in a data base environment, the data base software resides in the general-purpose computer. The software performs the functions of communicating with the end users or application programs, translating data base requests into RAP programs, transmitting these programs to RAP, controlling data base security and integrity, and performing data compression and decompression where needed.

Figure 16.3 illustrates the basic components of a cell. As a memory record passes the read head (R), it is copied into a buffer and sent to the microprocessor. If the microprocessor so chooses, it can modify parts of the record in the buffer. When the memory record passes the write head (W), the buffer is copied into the record area.

Relational Data Structures

Since RAP is oriented toward the relational data base concept, it is necessary to summarize a few ideas and terms associated with this concept. Using traditional terminology for the moment, one can think of

THE RELATIONAL ASSOCIATIVE PROCESSOR

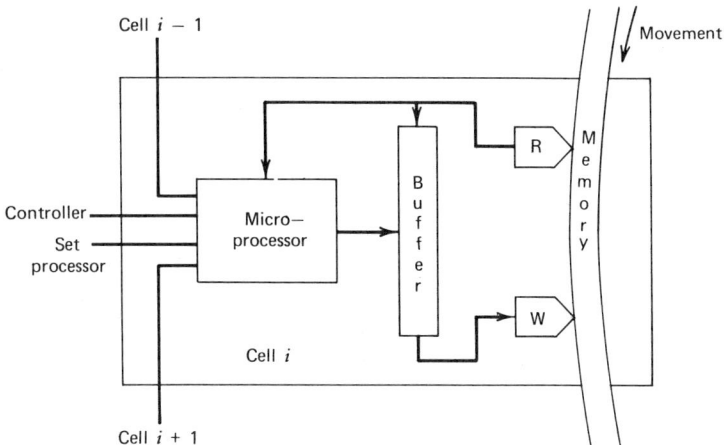

Figure 16.3 RAP cell structure.

a data base as a set of logical files, a logical file as a set of records, and a record as a sequence of fields. The records in a logical file are homogeneous; they have the same number of fields, and field n in one record has the same meaning and attributes as field n in every other record in the logical file.

In the relational concept, a *relation* corresponds to a logical file, a *tuple* corresponds to a record, a *domain* corresponds to a field, and a *domain name* corresponds to a field name, number, or position. In other words, a relation is viewed as a table of data about a set of similar entities. The name of the table is the relation name, each row is a tuple, an element in the table is a domain, and the column headings are the domain names. A relational data base is a set of relations that are interrelated through common domains.

In addition to this logical view of a data base, the relational concept defines a set of basic operations on the data base. Selection operations apply a boolean search predicate to a relation to select the tuples satisfying the predicate. The join operation is used to select tuples from a relation based on search criteria from another relation; the association is made through a domain name that is common to both relations. Set operations (e.g., union, intersection, complement, and difference) can be applied to subsets of a relation. The projection operation selects a subset of domains from a relation such that no duplicate domains appear. A free-variable operation selects tuples based on the domain values of other tuples. Finally, arithmetic and update operations on domains are present.

RAP Data Structures

The physical data representation in RAP is virtually the same as the relational data representation. Tuples in relations may appear anywhere within the rotating memories, but a single track can only contain tuples from the same relation.

The physical format of a track is shown in Figure 16.4. The first block on the track contains the name of the relation in which the tuples on the track reside. The relation name is encoded into 8, 16, or 32 bits. The second block contains the encoded names of the domains in the relation. Relation and domain names are represented in one of three item formats, as shown in the lower portion of Figure 16.4.

The remaining blocks contain tuples in the relation. The first bit (DF) is a delete flag; if it is on, the tuple is no longer needed and the space it occupies will eventually be reclaimed by garbage-collection logic in the cell. The mark bits correspond to the response array in Figure 16.1. A tuple is said to be T-marked (T is any combination of the 4 bits) if the corresponding marks bits are on. A tuple is T-unmarked if the corresponding mark bits are off.

The remaining fields in the tuple contain the values of the domains. The order of these values must be the same as the order of names in the domain-names block. Each value can be represented in 8, 16, or 32 bits, as indicated in Figure 16.4. The DL (delimiter) item is the last item in a block. Items that have a numerical value are stored in two's-complement binary notation.

A track must be preformatted before tuples are placed on it. A track is preformatted by writing the relation-name and domain-names blocks and then writing "empty" tuple blocks until the end of the track is encountered. The first 2 bits of the first item in the first tuple are set to 11 (TKE); TKE (logical track end) designates that this tuple block and all succeeding ones are empty (free space). If a tuple is later placed on this track, the tuple is written into the tuple block designated as TKE and the TKE designation is transferred to the succeeding tuple block. If a tuple is deleted, the garbage-collection logic eventually moves all tuples toward the front of the track so that the available space resides at the end of the track. When a new tuple is added, all the tracks corresponding to this relation are examined for available space. If none is found, a new cell is automatically allocated and its track preformatted.

As a note, the garbage-collection process seems to serve no purpose. Since the tracks are preformatted to the tuple size and the RAP operations have an execution time of one or more revolutions, there is no apparent advantage to moving the valid tuples to the front of the tracks.

THE RELATIONAL ASSOCIATIVE PROCESSOR

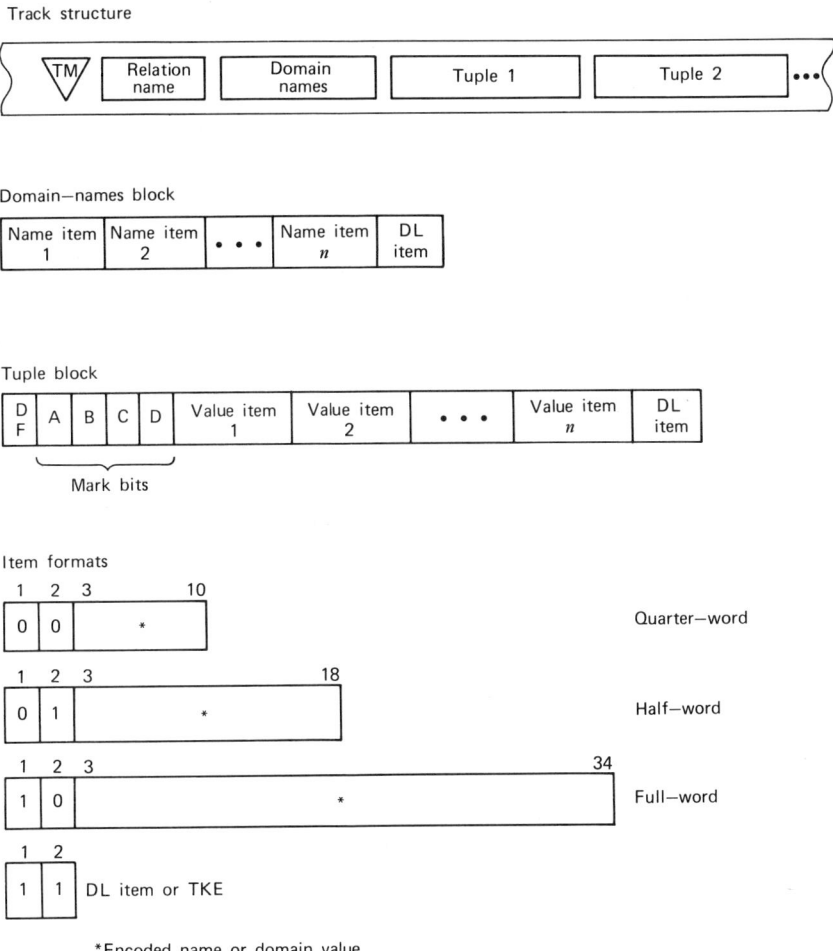

Figure 16.4 RAP data formats.

RAP Instruction Format

RAP contains a set of machine instructions that are used to form RAP programs to execute the relational data base operations. Rather than illustrating the instructions in machine-code form, they are illustrated in a symbolic assembly language.

The format differs somewhat from instruction to instruction, but the general form of a RAP instruction is

LABEL OPCODE (MARK BITS) [OBJECT: QUALIFICATION]

LABEL is an optional symbolic address; it is used to identify the targets of branch instructions. OPCODE designates the operation to be performed. The MARK-BITS field specifies a combination of mark bits to be set or reset by the instruction.

The OBJECT field specifies the relation name on which the instruction operates and sometimes the names of domains within that relation. It has the format RN or RN(DN1,DN2, . . . ,DNn) where RN is a relation name and DN1,DN2, . . . ,DNn are domain names.

The QUALIFICATION field is a boolean predicate of conditions that select the tuples to be associatively addressed. The predicate can be (1) null (implying that all tuples in the relation are being addressed), (2) a conjunction ("and") of conditions, or (3) a disjunction ("or") of conditions. Each condition in the predicate can have one of the forms:

1. RN.DN COMP OPERAND
2. RN.MKED(T)
3. RN.UNMKED(T)

The COMP field is a comparison operator (equal, not equal, less than, less than or equal, greater than, or greater than or equal). The OPERAND field can specify a RAP register, a literal value, or the name of a variable (i.e., an address) in the general-purpose computer. Nonnumeric literals are enclosed in quotation marks; variable names are enclosed in parentheses. The names of the addressable RAP registers are REGF1, REGF2, REGS, REGC1, REGC2, and REGC3. T is a specification of a combination of mark bits; valid values are A, B, C, D, AB, AC, AD, BC, BD, CD, ABC, ABD, ACD, BCD, and ABCD.

An example of a qualification is

(EMP.SALARY>5000) AND (EMP.MKED(A))

which addresses those tuples in relation EMP whose salary is greater than 5000 and whose mark bit A is on.

RAP Instruction Set

RAP contains 26 machine instructions out of which data base manipulation programs are constructed by the general-purpose computer. Since some of the instructions are rather complicated in function, none of the instructions is completely described; the instructions are described only to a level that gives the reader a feel for their functions. To present examples of some of the instructions, the data base in Figure 16.5 is used. Note that the two relations (EMP and ITEM) have a domain in common—DEPT.

THE RELATIONAL ASSOCIATIVE PROCESSOR

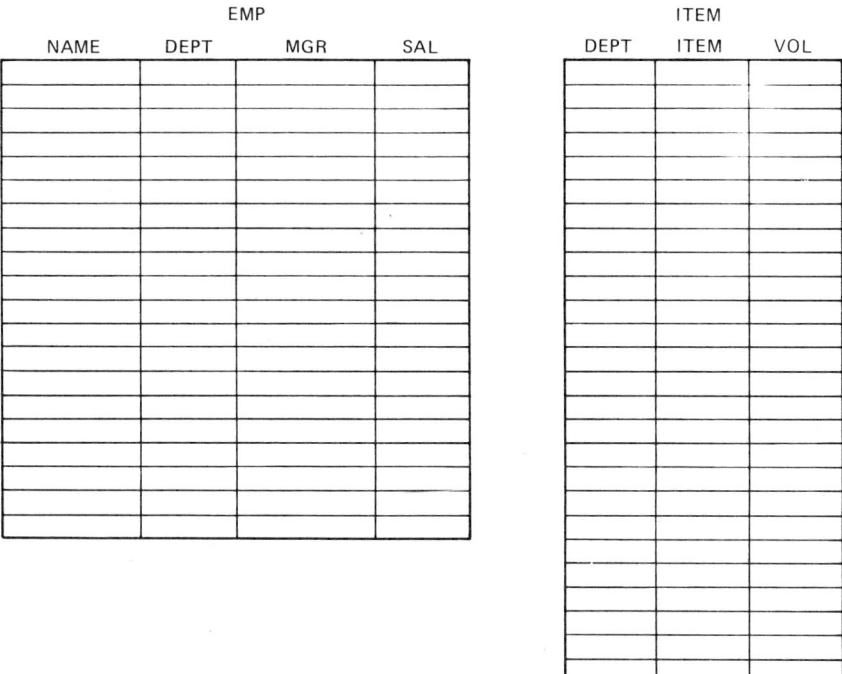

Figure 16.5 Example data base.

Instruction: MARK(T) [RN: QUALIFICATION]

Function: All tuples in relation RN meeting the qualification are marked with mark bits T.

Execution time: One revolution.

Example: The instruction MARK(AB) [EMP: EMP.DEPT='A42'] turns on marks A and B in the tuples representing employees in department A42.

Instruction: RESET(T) [RN: QUALIFICATION]

Function: Mark bits T are turned off (reset) in all tuples in relation RN meeting the qualification.

Execution time: One revolution.

Instruction: READ [OBJECT: QUALIFICATION] [WORKAREA]

Function: Data from the qualified tuples are transferred to the area specified by WORKAREA in the general-purpose computer.

Execution time: Depends on the amount of qualified data and the number of parallel I/O channels to RAP.

Example: The instruction READ [EMP: MGR='SMITH'] [BUF] transfers all EMP tuples whose manager is Smith to area BUF. The instruction READ [EMP(NAME,SAL): EMP.MKED(A)] [BUF] reads name and salary pairs of all employees whose mark A is currently on.

Instruction: READ-REG [register,register, . . .] [WORKAREA]

Function: The contents of the specified RAP registers are copied into the area specified by WORKAREA in the general-purpose computer.

Execution time: Less than one revolution.

Instruction: CROSS-MARK(T1) [RN1: RN1.DN1 COMP RN2.DN2: RN2.MKED(T2)]

Function: The join operation is performed by marking the tuples in the target relation (RN1) with respect to the COMP criterion between domain DN1 in the target relation and domain DN2 in the source relation (RN2). There may be several RN1 tuples for each RN2 tuple.

Execution time: $1 + NT2/K + N$ revolutions. NT2 is the number of T2-marked tuples in RN2, N is the number of tracks containing T2-marked tuples in RN2, and K is an implementation-dependent parameter.

Example: Given the instruction sequence
MARK(A) [ITEM: ITEM.VOL>100]
CROSS-MARK(A) [EMP: EMP.DEPT = ITEM.DEPT: ITEM.MKED (A)] the second instruction marks all employees working in departments that have sold more than 100 units of any item.

Instruction: CRS-COND-MARK(T1[T11]) [RN1: RN1.DN1 COMP RN2.DN2: RN2.MKED(T2)]

Function: Similar to that of CROSS-MARK. The T1 bits of a target tuple are reset (to 0) if the qualification criteria are not satisfied. T11 is a mark bit that is made available as a temporary area during the instruction. This instruction is used to perform an AND operation between two mappings on the same relation. It is normally preceded by a CROSS-MARK instruction.

Execution time: One more revolution than CROSS-MARK.

Example: CRS-COND-MARK is used when more than one relation is to be joined (cross-marked) into one target relation. Such an example is too complex to be shown here.

Instruction: GET-FIRST-MARK(T1) [RN(DN1,DN2, . . .): RN.DN1 COMP RN.DN2: RN.MKED(T2)]

THE RELATIONAL ASSOCIATIVE PROCESSOR

Function: Too complicated to be explained here. The instruction is used in performing the projection operation. As a note, the instruction places values in the REGC1, REGC2, and REGC3 registers.

Instruction: GET-FIRST [RN(DN1,DN2, . . .): RN.MKED(T)]
Function: A variation on GET-FIRST-MARK.

Instruction: SAVE [RN(DN): QUALIFICATION]
Function: The value of domain DN of a tuple meeting the qualification is placed in register REGS.
Execution time: One revolution.
Example: SAVE [ITEM(ITEM): ITEM.VOL < (MIN)] finds an item whose volume is less than the value of variable MIN and places the item number in REGS.

Instruction: ADD(T) [RN(DN): QUALIFICATION] [OPERAND]
Instruction: SUB(T) [RN(DN): QUALIFICATION] [OPERAND]
Instruction: MUL(T) [RN(DN): QUALIFICATION] [OPERAND]
Instruction: DIV(T) [RN(DN): QUALIFICATION] [OPERAND]
Function: The operation is performed on the DN domain of all qualified tuples. T is optional; if used it causes the qualified tuples to be T-marked as the operation is performed. This is intended for recovery purposes; if RAP fails during the instruction, the operation can be resumed on the unmarked tuples.
Execution time: One revolution.
Example: SUB [ITEM(VOL): (ITEM.DEPT='B46') AND (ITEM.VOL > 100)] [1] subtracts 1 from the volume domain in all items in department B46 with a volume of greater than 100.

Instruction: REPLACE(T) [RN(DN): QUALIFICATION] [OPERAND]
Function: The value OPERAND is stored in the DN domain of all qualified tuples. T is an optional specification of mark bits for recovery.
Execution time: One revolution.

Instruction: SUM [RN(DN): QUALIFICATION] [I]
Instruction: MAX [RN(DN): QUALIFICATION] [I]
Instruction: MIN [RN(DN): QUALIFICATION] [I]
Instruction: COUNT [RN: QUALIFICATION] [I]
Function: These are set-function instructions. The first three perform the operation on domain DN of all qualified tuples. COUNT counts the

number of qualified tuples. I has the value 1 or 2 and specifies which of registers REGF1 or REGF2 will hold the result.

Execution time: One revolution plus a fraction.

Example: The instruction SUM [ITEM(VOL)] [1] produces the sum of volumes of all tuples in ITEM. The instruction MAX [EMP (SAL): EMP.DEPT = 'A70'] [1] finds the maximum salary of an employee in department A70.

Instruction: DELETE [RN: QUALIFICATION]

Function: All qualified tuples are deleted (i.e., their delete flag is set).

Execution time: One revolution.

Instruction: INSERT [RN] [WORKAREA]

Function: The tuple in the area specified by WORKAREA in the general-purpose computer is added to the specified relation in the data base.

Execution time: One revolution if an existing relation track has available space. Two revolutions if a new track must be allocated and preformatted.

Instruction: DROP-DOMAIN [RN(DN): QUALIFICATION]

Function: The specified domain in all tuples in the relation is removed. All the tuples must have been previously marked; the mark is specified by QUALIFICATION.

Execution time: One revolution.

Instruction: DESTROY [RN] [CELL]

Function: If CELL is not specified, the entire relation is destroyed. If CELL (a cell number) is specified, only the tuples in that cell are removed from the relation.

Execution time: Fraction of a revolution.

Instruction: CREATE [RN] [WORKAREA] [CELL]

Function: A free cell is selected and preformatted for the specified new relation. If CELL is specified, this cell is selected. The specification says nothing about WORKAREA, but presumably it is an area in the general-purpose computer that contains the domain names and a description of the tuple.

Execution time: One revolution.

Instruction: TEST T-RAIL

Function: T is the character A, B, C, or D. If a T-mark bit is on in any tuple, a corresponding bit in the controller (called RAIL-STAT(T)) is set.

Execution time: Fraction of a revolution (each cell contains a 4-bit register; a bit in this register is on if the corresponding mark bit in any tuple in the cell is on).

Instruction: BC ADDRESS,CONDITION

Function: If the specified condition is true, control is transferred to the instruction whose label is specified by ADDRESS. CONDITION can be
1. RAIL-STAT(T)
2. REGISTER COMP OPERAND
3. Null (an unconditional branch)

Execution time: Fraction of a revolution.

Example: The instruction BC LOOP,REGS='A40' transfers control to the instruction labeled LOOP if register REGS has the value A40. The instruction BC L1 transfers control to the instruction labeled L1.

Instruction: EOQ

Function: This instruction terminates a RAP program.

Execution time: Negligible.

RAP Program Examples

To illustrate programs executed by RAP, two examples are given. The first program is the result of a data base request to delete from EMP all the employees who work for J. Jones' manager. The RAP program is

    SAVE [EMP(MGR): EMP.NAME='J.JONES']
    DELETE [EMP: EMP.MGR=REGS]
    EOQ

The execution time is slightly more than two revolutions (plus, on the average, an additional 0.5 revolutions of rotational delay).

The second program is the result of a request to count the distinct number of items and the volume of items sold in the department in which B. Smith is employed. The RAP program is

1. MARK(A) [EMP: EMP.NAME='B.SMITH']
2. CROSS-MARK(A) [ITEM: ITEM.DEPT=EMP.DEPT: EMP.MKED(A)]
3. COUNT [ITEM: ITEM.MKED(A)] [1]

4. SUM [ITEM(VOL): ITEM. MKED(A)][2]
5. READ-REG [REGF1,REGF2] [BUF]
6. RESET(A) [ITEM]
7. RESET(A) [EMP]
8. EOQ

Instruction 1 marks the EMP tuple for B. Smith. Instruction 2 marks the ITEM tuples associated with B. Smith's department. Instructions 3 and 4 place the count of marked items and the sum of the volume of the marked items into registers REGF1 and REGF2. Instruction 5 reads these values into area BUF in the general-purpose computer. The RESET instructions reset the mark bits that were used.

A RAP Virtual Memory

The size of an economically feasible RAP has been estimated to be approximately one million bits. However, because many actual data bases are larger than this, an extension to RAP incorporating the concept of a virtual memory has been proposed [17].

To briefly summarize this extension, the entire data base is stored in nonassociative bulk memory attached to the general-purpose computer. Each cell in RAP contains two memory components; at any one point in time, one is designated as the active memory and the other as the buffer memory. The general-purpose computer has access to the current buffer memories via a separate I/O controller and channel.

When the general-purpose computer builds a RAP program, it determines if the necessary relations are stored in RAP. If they are not, it moves them from the bulk memory to the RAP buffer memories (look-ahead paging) while RAP is perhaps processing another program. The page-replacement algorithm is to look first for buffer memories that have not been updated and are not needed by programs in the RAP queue. If the only buffer memories available contain relations that have been altered, they are first paged back to the bulk memory. Once the required pages have been transferred, the program is transmitted to RAP along with directions about which memory pairs to swap (buffer to active, active to buffer).

This virtual memory approach is based on an intuitive, but untested, feeling that data base requests exhibit a "locality of reference" similar to the locality-of-reference phenomenon that is well known in existing main-storage virtual memories. Some of the intuitive arguments follow. As a deadline approaches (e.g., an airlines flight), activity associated with certain relations (e.g., a relation for that flight) should increase substantially, thus increasing the locality of reference within the data base. Also, the use of interactive query languages encourages browsing (searching for relevant data in a process of stepwise refinement), causing a set of relations to be accessed frequently over a short period of time. Finally, batch-updating or report-writing programs tend to reference a set of relations iteratively, again increasing the locality of reference within the data base.

THE ONE-LEVEL STORE

An extremely important concept in computer architecture is the idea of a *one-level store*. This concept is based on the notion that "storage is storage is storage" and that "storage management is storage management is storage management." That is, why design two or more dissimilar interfaces to reference data in storage (i.e., load, move, and add instructions if the data are in main storage, I/O instructions if the data are in secondary storage)? Why burden the application program with the problems of managing a storage hierarchy (e.g., by explicitly using GETs and PUTs to move data to and from main storage)? Why manage a virtual main storage (e.g., via paging and/or segmentation) in a way distinct from the management of secondary storage (e.g., via "access methods")?

An implementation of part of the one-level-store concept exists in the Multics system, although the concept is implemented at the operating system, rather than computer architecture, level [20,21]. Multics dispenses with the idea of files as a separate mechanism to address data. This is accomplished by unifying the mechanisms of main-storage and file management; a program's files are represented as part of the program's address space. There is no need for a program to issue read or write I/O requests to move data between a file and main storage. A program addresses words within a file as if the file were part of its virtual memory (i.e., a word is referenced by a segment-name, word-number couple). Virtual-storage management in Multics treats file accesses in the same way it treats paging operations to a paging disk.

The main advantage of the one-level-store concept is that it simplifies and unifies an important aspect of any architecture: the concept of storage. This is illustrated in the following section. This advantage also reduces the complexity of both the system software and the application programs. Finally, the one-level store provides total I/O device independence by completely shielding programs from the characteristics of the actual storage devices.

The concept of a one-level store usually implies that the underlying machine is moving data and programs through a hierarchy of storage devices (e.g., high-speed cache, slower-speed semiconductor random-access memory, fixed-head disks, drums, or magnetic bubbles, moving-head disks, and tape or laser mass storages). Madnick, for one, discusses several current issues in managing such a hierarchy [22].

I/O IN THE SWARD MACHINE

The SWARD architecture discussed in Chapters 13–15 had no provisions for input/output. However, it might be instructive to consider how these provisions might be added to the architecture.

The first consideration is that I/O can be classified in two ways: *memory* and *source/sink*. Memory I/O is associated with devices that have a memory and are usually randomly addressable; this category includes files and data bases stored on such media as rotating disks and drums. Source/sink I/O is characteristically memoryless and purely sequential (i.e., it behaves like a first-in/first-out queue); this category includes such devices as card readers, terminals, and communication lines. Magnetic tapes appear to exhibit some characteristics of both categories; however, they are usually considered to be in the source/sink category.

The memory I/O category is considered first. If one employs the one-level-store concept, the architecture is seen to have memory I/O with virtually no changes in the definition of the architecture. That is, one can think of a file as a one-dimensional array of structures. Each array element corresponds to a record in the file. The nested tag in the array cell would likely be a structure, where the structure defines the fields in each record. Since the existing machine instructions are generic and apply to arrays and array elements, the "I/O instructions" are the existing instructions.

One change that is required is a change to the ALLOCATE instruction, since all storage objects are currently deleted when the program terminates. A one-token immediate field might be added to the instruction. This field would indicate whether the object is to be deleted at program termination. To create a permanent "file," a program issues the ALLOCATE instruction, indicating that the created storage object should not be freed on program termination. The ALLOCATE instruction points to a relocatable array which in turn points to a pointer cell. The logical address (capability) that is returned serves to uniquely identify the file until it is deleted (with a FREE instruction). The file is constructed by executing MOVE instructions to move data into the array elements.

Since logical addresses appear, to a human, to be a meaningless series of bits, an instruction might be added to allow the program to communicate with the operating system. In this case, the program might use this instruction to inform the operating system to "associate this symbolic name with this logical address." (In a moment, it is shown that this new instruction is unnecessary.)

To retrieve an existing file, a program simply refers to a relocatable array based on a pointer cell containing the unique logical address of the object (the file). Rather than having people carry logical addresses from program to program, the operating-system instruction could be used to say "Here is a symbolic name and a pointer cell; from your directory, put the associated logical address in the pointer cell."

Another consideration is that array cells can have a maximum of 65,535 elements in each dimension (an arbitrary limit). One might redefine the array cell so that the upper-bound fields are only 3 tokens in length, calling this a "small array" cell. Another array cell might be introduced with 6-token upper-bound fields (an upper limit of about 17 million elements in each dimension), calling this a "large array" cell.

Given the removal, at the architectural level, of the distinction between main-memory operations and secondary-storage I/O, a natural extension is to carry this notion into programming languages, that is, the removal of file-I/O statements from programming languages.

One problem associated with a one-level store as described above that deserves more research is the mechanism with which a program searches a file (represented as an array) to locate a particular record (array element). If hash addressing can be used, it is natural to represent files as arrays. However, if hash addressing is inappropriate for a particular file, the only other solution appears to be an iterative sequential search (unless the file is ordered by the search field, implying that a binary search could be used), which is unsatisfactory for large files. Hence the possibility of storing one or more indexes with array cells comes to mind. Another possibility is allowing designated arrays to be content addressable. In short, the relationships between the concepts of one-level stores and data base processing need further investigation.

With respect to source/sink I/O, the SWARD architecture could be extended in a similar way, again without the need for explicit I/O instructions. Let us assume that every source/sink device (e.g., printer, card reader, magnetic tape, terminal, and communication path to another system or program) is identified with a unique reserved logical address (capability). Think of a relocatable cell based on a pointer cell containing one of these reserved logical addresses as representing the top or bottom of a queue; the queue represents the source/sink device. For instance, to read from an 80-character card reader, the program defines a relocatable character-string cell of length 80. The associated pointer cell contains the logical address of the particular source device. When the program refers to the character-string cell as the target operand of a MOVE instruction, the underlying machine first obtains the next card and establishes it as the value of the relocatable character string. The end-of-file condition would be represented as the undefined value. The program can explicitly test for this using the DEFINED instruction or allow the existing fault-handling mechanism to handle it (undefined-operand fault).

Sink devices would be handled in the same way, except that a move into a cell representing the device would cause the underlying machine to write the data on the device.

Note that the relocatable cell need not be associated with an I/O device. The cell could be asociated with another program, providing a simple and general send/receive mechanism to pass messages among programs. In fact, one of these programs could be the operating system, removing the need for the special instruction mentioned earlier to provide communication with the operating system. Assume that a standard is established stating that the logical address 111...111 will always represent the operating system. In addition, assume that the rules on pointer cells are slightly relaxed; programs can fabricate a logical address, but only the address 111...111. A program sends a message to the operating system by storing into a relocatable cell based on a pointer cell with this value. The operating system receives messages by moving data out of a relocatable cell based on a pointer cell holding logical address 111...111. Since, for reasons of reliability and security, the operating system should not have addressability to the program's address spaces, a program that sends a message to the operating system should include a logical address of a return area in the message. (Alternatively, the operating system could return results by sending messages to the program.)

To achieve device independence, one does not want to make programs aware of the special reserved logical addresses for sources and sinks. Hence, as described earlier, a program that is about to use such a device probably has a need to send the operating system the message "Here is a symbolic name for a source or sink and a pointer cell; put the associated logical address in the pointer cell."

In short, it appears that I/O concepts can be blended into the SWARD architecture by simply extending the concept of capabilities or logical addresses to files, devices, and programs; no explicit I/O instructions are needed.

REFERENCES

1. E. I. Lowenthal, "Backend Machines for Data Base Management: A Tutorial," *Proceedings of the Fifth Texas Conference on Computing Systems*. New York: IEEE, 1976, pp. 21–25.
2. R. H. Canaday et al., "A Back-end Computer for Data Base Management," *Communications of the ACM*, **17**(10), 575–582 (1974).
3. D. R. Anderson, "Data Base Processor Technology," *Proceedings of the 1976 National Computer Conference*. Montvale, N.J.: AFIPS, 1976, pp. 811–818.
4. G. S. Christensen and P. D. Jones, "The Control Data Star-100 File Storage Station," *Proceedings of the 1972 Fall Joint Computer Conference*. Montvale, N.J.: AFIPS, 1972, pp. 561–569.
5. J. A. Rudolph, "A Production Implementation of an Associative Array Processor—STARAN," *Proceedings of the 1972 Fall Joint Computer Conference*. Montvale, N.J.: AFIPS, 1972, pp. 229–241.
6. G. A. Anderson and R. Y. Kain, "A Content-Addressed Memory Designed for Data Base Applications," *Proceedings of the 1976 International Conference on Parallel Processing*. New York: IEEE, 1976, pp. 191–195.
7. R. I. Baum, D. K. Hsiao, and K. Kannon, "The Architecture of a Database Computer. Part I. Concepts and Capabilities," OSU-CISRC-TR-76-1, Ohio State University, 1976.
8. L. D. Healy, G. J. Lipovski, and K. L. Daty, "The Architecture of a Context Addressed Segment-Sequential Storage," *Proceedings of the 1972 Fall Joint Computer Conference*. Montvale, N.J.: AFIPS, 1972, pp. 691–701.
9. B. Parhami, "A Highly Parallel Computing System for Information Retrieval," *Proceedings of the 1972 Fall Joint Computer Conference*. Montvale, N.J.: AFIPS, 1972, pp. 681–690.
10. C. S. Lin, D. C. P. Smith, and J. M. Smith, "The Design of a Rotating Associative Memory for Relational Database Applications," *ACM Transactions on Database Systems*, **1**(1), 53–65 (1976).
11. E. F. Codd, "A Relational Model of Data for Large Shared Data Banks," *Communications of the ACM*, **13**(6), 377–387 (1970).
12. C. J. Date, *An Introduction to Database Systems*. Reading, Mass.: Addison-Wesley, 1975.
13. E. A. Ozkarahan, S. A. Schuster, and K. C. Smith, "An Associative Processor for Data Base Management," *Proceedings of the 1975 National Computer Conference*. Montvale, N.J.: AFIPS, 1975, pp. 379–387.
14. C. R. DeFiore and P. B. Berra, "A Quantitative Analysis of the Utilization of Associative Memories in Data Management," *IEEE Transactions on Computers*, **C-23**(2), 121–133 (1974).
15. G. G. Langdon, Jr., "A Note on Associative Processors for Data Management," RJ-1941, IBM Research Laboratory, San Jose, Calif., 1977.
16. E. A. Ozkarahan, S. A. Schuster, and K. C. Smith, "A Data Base Processor," CSRG-43, University of Toronto, 1974.
17. S. A. Schuster, E. A. Ozkarahan, and K. C. Smith, "A Virtual Memory System for a Relational Associative Processor," *Proceedings of the 1976 National Computer Conference*. Montvale, N.J.: AFIPS, 1976, pp. 855–862.

18. S. A. Schuster, E. A. Ozkarahan, and K. C. Smith, "The Case for a Parallel-Associative Approach to Data Base Machine Architectures," *Proceedings of the Berkeley Workshop on Distributed Data Management and Computer Networks*. Springfield, Va.: National Technical Information Service, 1976, pp. 365–375.
19. E. A. Ozkarahan, S. A. Schuster, and K. C. Sevcik, "Performance Evaluation of a Relational Associative Processor," *ACM Transactions on Database Systems*, 2(2), 175–195 (1977).
20. R. C. Daley and J. B. Dennis, "Virtual Memory, Processes, and Sharing in MULTICS," *Communications of the ACM*, 11(5), 306–312 (1968).
21. E. I. Organick, *The Multics System: An Examination of its Structure*. Cambridge, Mass.: MIT Press, 1972.
22. S. E. Madnick, "INFOPLEX—Hierarchical Decomposition of a Large Information Management System using a Microprocessor Complex," *Proceedings of the 1975 National Computer Conference*. Montvale, N.J.: AFIPS, 1975, pp. 581–586.

EXERCISES

16.1 Given the data base in Figure 16.5, write a RAP program to add $100 to the salary of all employees in the department(s) associated with the item(s) of largest volume. However, do not increase the salary of employees whose salary is $20,000 or more.

16.2 Given that the associative memory completely circulates (rotates) every 20 milliseconds, what is the expected execution time of this program? (Assume that there are eight employees in the qualified department(s) and that K=4.)

16.3 The obvious factor that makes an associative store such as RAP outperform a conventional data-management system is its parallelism. However, there are other significant factors in a RAP-like subsystem that lead to increased performance. What are they?

16.4 What is the other significant advantage, in addition to the performance advantage, of an associative store?

16.5 Two files in the SWARD machine are represented by arrays with the symbolic names A and B. Write a sequence of assembly-language instructions to copy file B into file A.

17 | Architecture Optimization and Tuning

Given an initial design of a computer architecture, an important, but often overlooked, consideration is the optimization of the architecture using concepts of information theory and a variety of heuristics. The optimization of an architecture is a process of attempting to minimize the number of bits needed to represent programs and data and minimizing the number of bits that must be transferred between the processor and memory to execute a given program. The motivations are (1) reducing the memory space needed, (2) reducing execution time by making more effective use of the available bandwidth between the processor and memory, and, as a valuable side effect, (3) removing arbitrary restrictions on the upper limit of such entities as operation codes and addresses.

There are two basic approaches to optimization: (1) optimizing the instruction set (e.g., by adding new instructions) and (2) optimizing the representation and size of instructions, data, and addresses, although, as seen later, the two approaches are theoretically the same. An important consideration is that detailed information about the characteristics of programs is necessary to perform the optimization processes. This information must include the relative frequencies of machine instructions, the relative frequencies of sequences of instructions (serial dependencies, e.g., the probability that a load instruction is followed by an add instruction), and the relative frequencies of data and address values. At first it appears that the only way of obtaining this data is to build the machine, compile programs to it, and then measure the attributes of these programs. However, it is shown later that these data can be obtained without the necessity of building the machine and its compilers.

A potential problem is that the information can be obtained in two ways: *static* or *dynamic* frequencies. For instance, one way of obtaining instruction-frequency data is to count the instructions as they reside in storage or as they are generated by a compiler. These are static frequencies. Presumably, if one uses this information for optimization, the optimization is oriented toward storage space but not necessarily execution time (except that minimizing storage space tends to reduce a program's working-set size, which tends to reduce execution time). Alternatively, one could obtain instruction frequencies by counting instructions as they are executed, a dynamic frequency measurement. Using dynamic frequencies, the resultant optimizations appear to be oriented toward execution time, not storage space.

This dilemma suggests that we must decide whether to optimize for either space or time, but not both. Fortunately, however, the dilemma rarely arises. Data from many sources indicate a close correlation between static and dynamic frequencies; thus we can select one or the other and be reasonably assured that we will be optimizing for both space and time. As evidence, Table 17.1 illustrates the eight most frequent instructions, measured statically and dynamically, in 19 programs on the IBM S/360 [1].

INSTRUCTION-SET OPTIMIZATIONS

One type of optimization is an analysis of the use of machine instructions with the goal of creating new instructions to save time and space. The principal heuristic is to measure the frequencies of sequences of

Table 17.1 Most Frequent S/360 Instructions

Static (%)		Dynamic (%)	
L	28.6	L	27.3
ST	15.0	BC	13.7
BC	10.0	ST	9.8
LA	7.0	C	6.2
SR	5.8	LA	6.1
BAL	5.3	SR	4.5
SLL	3.6	IC	4.1
IC	3.2	A	3.7

INSTRUCTION-SET OPTIMIZATIONS

pairs or triplets of instructions, creating new instructions that have the combined functions of high frequency sequences. As a simple example, the instruction pair

```
SR  register,register
IC  register,address
```

was found to be a high-frequency pair in the S/360 (representing 2.7% and 3.8% of all static and dynamic instruction pairs) [1]. The SR (subtract-register) instruction clears a register, and the IC (insert-character) instruction places a character value in the cleared register. The improvement would be the creation of a "clear register and insert character" instruction. This instruction occupies less space and has a faster execution time than the SR-IC pair. This trivial improvement could be expected to reduce the total average program size by 1% and the total average execution time of a program by 1.5%.

As a second example, a study of programs for the IBM S/370 indicated that the following sequence of instructions was frequently generated by a compiler [2]:

```
ST  REG,SAVE
L   REG,VARIABLE
LA  REG,N(,REG)
ST  REG,VARIABLE
L   REG,SAVE
```

The purpose of the instructions is to increment a binary-integer variable by N. Since, in this machine, binary arithmetic can only be performed in registers and the registers are a scarce commodity, the two load and store pairs are needed. On a S/370 Model 145, these instructions have a total execution time of 8.8 microseconds and occupy 20 bytes of storage. If a new instruction called INCREMENT was available in the form

```
INC  VARIABLE,INCREMENT-VALUE
```

the five instructions could be replaced with a single instruction occupying 4 bytes of storage with an estimated execution time of 2.7 microseconds. (The astute reader may disagree, saying that a LA instruction can add an increment of up to 4095 but the INC instruction can add an increment of only up to 16; however, a frequency study of increment values would show that the INC instruction is usable in the vast majority of situations.)

Student-PL Machine Optimization

The primary purpose of the Student-PL machine (Chapters 5–7) was to serve as a vehicle for studying techniques of architecture optimization [3]. Chapters 5–7 presented the unoptimized version of the machine; here we consider the improvements in the architecture.

The architecture was optimized by studying the characteristics of 959 programs compiled to the unoptimized version; the programs contained a total of 37,443 Student-PL statements. The effects of each optimization change were quantified in terms of the factors:

A1 number of instruction bits in a program.
A2 number of data bits in a program.
A3 number of memory references for instructions.
A4 number of memory references for data.
A5 number of instruction bits fetched for a program.
A6 number of data bits fetched for a program.

The first step in the analysis was to measure instruction frequencies. The dynamic and static frequencies of the five most frequent instructions are shown in Table 17.2. NAME represents the SNAME and LNAME instructions.

The first improvement was the elimination of the LINE instruction. Since the correspondence between machine instructions and programming-language statement numbers is static, there is no need to continually reflect this correspondence during the dynamics of program execution; an alternative is to use a software-constructed correspondence table in secondary storage. This simple change reduced A1 by an average of 14%, A3 by 8.7%, and A5 by 12.2%.

The next step was a study of the frequency of instruction pairs. The static frequency of the six most frequent pairs is shown in Table 17.3 (the discarded LINE instruction was omitted from this study).

Table 17.2 Most Frequent SPLM Instructions

Instruction	Static (%)	Dynamic (%)
NAME	30.3	28.3
EVAL	20.7	23.7
LINE	10.0	8.7
SWAP	4.8	4.1
POP	4.8	2.7

INSTRUCTION-SET OPTIMIZATIONS

Table 17.3 Most Frequent SPLM Static Instruction Pairs.

Instruction Pair	Frequency (%)
NAME, EVAL	19.8
NAME, NAME	8.9
EVAL, NAME	6.5
PARAM, SWAP	4.1
POP, NAME	3.6
STORE, POP	3.3

The extremely frequent NAME/EVAL pair is used to bring a value to the top of the stack. The obvious improvement is to add an instruction to perform the combined function, namely, SLOAD (short load, corresponding to SNAME and EVAL) and LLOAD (corresponding to LNAME and EVAL). Hence instead of a compiler generating the instructions

SNAME 1,3
EVAL

to bring the value of the variable at address 1,3 to the top of the stack, it generates

SLOAD 1,3

The addition of SLOAD and LLOAD reduces factor A1 by an average of 23%, as well as reducing A3 and A5. Note that the load instructions are additions to the instruction set; the SNAME, LNAME, and EVAL instructions are not removed from the instruction set, since there are situations in which they are needed individually.

Four other similar changes were made. It was found that 72% of the STORE instructions were followed by a POP instruction; a new instruction (STORED) with the combined function of STORE and POP was introduced. Also, 71% of the SUBS instructions were followed by an EVAL instruction; thus the instruction SUBSEVAL was added. Further analysis of instruction pairs showed that every DOTEST instruction is followed by a CRET instruction, and every DOINCR instruction is followed by a CYCLE instruction. As a result, DOTEST and DOINCR were redefined to include the CRET and CYCLE operations.

The next improvement was in the area of addressing. Of the NAME instructions (SNAME and LNAME), 72.7% were found to be SNAME. The motivations for optimization were to increase this percentage and also decrease the lengths of SNAME and LNAME. The first step was a frequency analysis of lexical-level addresses. Two results of this analysis were (1) over 80% of the addresses in instructions refer to the outermost lexical level (lexical-level 1, since lexical-level 0 was used for system software) and (2) over 90% of the addresses in instructions in a given lexical level were for the same lexical level (i.e., the majority of references were to variables local to the current lexical level).

For these reasons, and because the 8-bit op-code field was not completely utilized (there are less than 256 instruction types), four changes were made. The SNAME instruction was reduced to 8 bits in length. Rather than using all 8 bits as an op-code in all instructions, if the first two bits are 00, the instruction is SNAME and the remaining 6 bits are defined as follows. If the SNAME instruction resides in lexical level 1, the addressed lexical level is assumed to be 1 and the 6 bits are an order number (0–63). If the instruction does not reside in lexical-level 1, the first bit designates a lexical level (0 = current lexical level, 1 = lexical-level 1) and the remaining 5 bits specify the order number. If the reader feels uncomfortable with this, the examples in Exercise 17.1 should be studied.

The second change was a reduction in the length of the new SLOAD instruction to 8 bits. This was done in an identical manner, except that the first 2 bits of an SLOAD instruction are 01.

The last two addressing changes were defining LNAME and LLOAD as 16-bit instructions in the following manner. The first 5 bits uniquely identify the instruction, the next 3 bits specify a lexical level, and the last 8 bits specify an order number. The effect of these four changes was to reduce A1 (total instruction bits) by an average of 23% and A5 (instructions bits fetched) by 13.5%. Also, it was observed that 92% of the addresses in an average program could now be represented in the short form (SNAME and SLOAD).

The next area of study was the representation of data, particularly the representation of constants (e.g., the constant 1 in the statement A=A+1). These observations about constants were made. First, constants are represented as words in the data stack and are addressed by lexical-level addresses. Since order numbers must be assigned to constants, they tend to increase the necessity of using the long forms of lexical-level addresses. Second, when a constant is needed, it must be fetched from the data stack, increasing factor A4. Finally, a frequency analysis of constants showed that 72% of them were FIXED (binary integers). Of these, half had the value 1, and 99% had a value less than 64. Hence representing constants as data-stack words leads to a waste of space (i.e., a large number of leading zeros), increasing factor A2.

Based on these observations, a new 8-bit instruction, LITERAL, was added. The first 2 bits have the value 10, which uniquely identifies this as a LITERAL instruction. The next 6 bits contain a value interpreted to be a positive binary-integer value to be loaded onto the top of the stack. As a result, only a small fraction of constants must be represented in the data stack. This reduces factors A2, A4, and A6. It also indirectly reduces factors A1 and A5 in that the reduction in order numbers decreases the necessity for long-form addresses (the addition of the LITERAL instruction was found to reduce the number of order numbers in each lexical level by an average of 24%). It also results in smaller symbol tables and less data-stack initialization time in the ENTER instruction.

The end result of these and a few other improvements was quite dramatic. On the average, the number of instruction bits in a typical program (A1) was reduced by 51%, the number of memory references for instructions (A3) was reduced by 37%, and the number of instruction bits fetched (A5) was reduced by 58%. Also, the number of memory references for data (A4) decreased by 50%, and the number of data bits fetched (A6) decreased by 62%. The average total size of a program was reduced by 23%, and the number of instructions executed decreased by 46%. An experimental comparison of Student-PL programs on the optimized SPLM versus equivalent PL/I programs on the IBM S/360 shows that the SPLM programs occupy 13 times less space and require 3.5 times fewer bits to pass across the memory-processor interface [3].

OPERATION-CODE OPTIMIZATION

Although it might not be apparent at this point, all the informal improvements in the previous section are based on the removal of *redundancy* in storage. The field of information theory [4] can be drawn on because it formalizes and quantifies such concepts as redundancy and information and provides tools to minimize redundancy.

One result from information theory is a theoretical measure of the actual information content of a message. The measure is often called the information rate of the source, the source entropy, the source uncertainty, or **H**. One can view the entire object program, the sequence of machine instructions, the program's data, the addresses in machine instructions, or the op-codes as the parts of a message. Storage itself, or the processor-memory path, can be viewed as a communication channel. If we view the parts of whatever message we are interested in as independent from one another (we have already seen that this is untrue for a sequence of machine instructions or op-codes, but let us assume that we wish to ignore this by dealing with only local optimizations), then **H** is given by

$$H = - \Sigma P_i \log P_i$$

P_i is the probability of occurrence of the ith message symbol. Since we are representing information in bits, all logarithms are to the base two. The redundancy of a particular encoding of the message symbols is given by

Redundancy = 1 − **H**/actual average symbol size

which ranges from zero (no redundancy) to 100% (infinite redundancy).

If these measures are used to optimize the representations of op-codes, P_i is the probability of the ith op-code, and the summation is over the total number of distinct op-codes. **H** is a measure of the average number of bits of information in each op-code. The probabilities of occurrence can be static (implying that we are interested in optimizing the space occupied by op-codes) or dynamic (implying that we are interested in optimizing the processor-memory data transfer). However, as noted earlier, there is a high correlation between static and dynamic frequencies of op-codes, meaning that we can arbitrarily choose one and rest assured that we are optimizing both factors.

As a simple example, consider a machine with seven instructions named A, B, C, D, E, F, and G. If we select a fixed-size op-code field for the machine, its optimal length is 3 bits. Hence a program of 1000 instructions will have 3000 op-code bits. Assume that the frequencies of the seven instructions are those shown in Table 17.4. The frequencies in Table 17.4 should not be viewed with surprise; we have already seen that instruction frequencies are rather skewed.

The **H** calculation on these instructions (treating the instructions as independent, that is, not considering the frequencies of sequences of instructions) is 1.88. Although the op-code length is 3 bits, the number of bits of information in each op-code is only 1.88. The redundancy is 1-(1.88/3) or 37%. This leads to the suspicion that there is a better representation of the op-codes.

The apparent optimization is to use a variable-size op-code field; the most frequent instructions will have the shortest op-code, and the least-frequent instructions will have the largest op-codes. Rather than

Table 17.4 Instruction Frequencies

Instruction	P_i
A	0.50
B	0.30
C	0.10
D	0.03
E	0.03
F	0.02
G	0.02

creating these op-codes in an ad hoc manner, there is an algorithm called *Huffman encoding* that will generate the optimal representation of op-codes [4]. In fact, a fundamental theorem from information theory shows that if we use Huffman encoding, the average op-code size in our case will be between 1.88 and 2.21 bits (2.21 is 1.88 + 1/3; the 3 is the block size of the symbols being encoded—the original 3-bit op-code).

Table 17.5 shows the Huffman-encoded op-codes. Notice that the most frequent instruction (A) has a 1-bit op-code and the least-frequent instructions have 5-bit op-codes. The average op-code length is obtained by summing the frequency of each instruction multiplied by its op-code length. The average op-code length is found to be 1.90 bits, the redundancy is only 1%, and the 1000-instruction program now has 1900, rather than 3000, op-code bits.

Notice that the op-codes are *uniquely decipherable*. That is, they were defined in such a way that the processor can examine an instruction and unambiguously determine the instruction type. If the op-code for B, for instance, was changed from 01 to 10, this property would not exist; the processor would not be able to distinguish between instructions A and B.

Table 17.5 Huffman-encoded Op-codes

Instruction	P_i	Op-code	Op-code Length
A	0.50	1	1
B	0.30	01	2
C	0.10	001	3
D	0.03	00000	5
E	0.03	00001	5
F	0.02	00010	5
G	0.02	00011	5

Although the Huffman-encoded op-codes represent the optimal representation, this representation is almost never used because it is impractical. It implies that memory must be addressable to the bit level, and, because each op-code could potentially have a different length, it implies a costly op-code decoding process in the processor. As a result, a compromise is usually made; two or more fixed sizes are chosen for op-codes, and the most-frequent instructions are given the smallest op-codes. As shown in several examples below, these compromises yield considerable savings over a single-size op-code.

For the hypothetical architecture, assume that op-codes can be either 2 or 4 bits in length. Table 17.6 illustrates the op-code assignments. Again, the op-codes have been defined such that they are uniquely decipherable. The average op-code length is 2.20 bits, the redundancy is 15%, and the 1000-instruction program has 2200, rather than 3000, op-code bits.

Another advantage of this type of op-code encoding not considered in the example is that it places no upper bound on the number of op-codes. In Table 17.6, the value 00 in the first 2 bits is an "escape code"; it indicates that the op-code is greater than 2 bits long. By carrying this idea to successive pairs of op-code bits, we can define an unlimited number of op-codes. However, to do this, the value 00...00 can never be used to represent an op-code. Hence in Table 17.6 G would be assigned the 6-bit op-code 000011. This increases the average op-code size to 2.24 bits, a small price to pay for this flexibility. In most situations, however, there would be no increase in average op-code size, since the number of instructions is rarely equal to all possible op-code values.

Op-Code Optimization in the B1700

In Chapter 12 we saw this type of optimization in the COBOL-oriented architecture of the Burroughs B1700. Op-codes are represented in 3 or 9

Table 17.6 Compromise Op-codes

Instruction	P_i	Op-code	Op-code Length
A	0.50	11	2
B	0.30	10	2
C	0.10	01	2
D	0.03	0011	4
E	0.03	0010	4
F	0.02	0001	4
G	0.02	0000	4

OPERATION-CODE OPTIMIZATION

bits. The seven most frequent instructions have 3-bit op-codes; the eighth value indicates that the next 6 bits are also part of the op-code. Here we consider the optimization of another architecture in the B1700: the SDL-oriented architecture [5].

The operating system and compilers in the B1700 are written in the SDL language and executed by the SDL-oriented architecture. When defining the SDL-oriented architecture, the designers examined the benefits of op-code encoding and decided on three lengths: 4, 6, and 10 bits. Of the 16 values of the first 4 bits of an instruction, 10 name the 10 most frequent instructions, five indicate that the op-code is 6 bits long, and the last indicates that the op-code is 10 bits long.

To study the desirability of this choice, it was compared to a fixed-length 8-bit representation on one extreme and the Huffman encoding on the other extreme. Table 17.7 compares these three representations based on the total number of op-code bits in the operating system and the relative penalty in op-code decoding time. As shown, the method selected proved to be a wise choice; it is almost as good as the Huffman encoding, but requires a relatively small decoding penalty. The 4-6-10 method reduced the number of op-code bits in the operating system by 39%. This proved to be significant, because op-codes in the SDL-oriented architecture occupy close to one-third of the total program space.

Op-Code Optimization in the SWARD Machine

Chapter 15, the definition of the instruction set of the SWARD machine, defined the op-code of each instruction as a variable multiple of 4 bits. The rationale for this frequency-based encoding is now discussed.

Since the op-codes were encoded when the machine was being specified on paper, the obvious question is, how were the instruction frequencies obtained? The data were obtained by the "language-fragment method" [3]. That is, data were obtained (e.g., from [1,3,7]) on the frequency of use of programming-language statements and operators, an analysis was made of which SWARD instructions would be generated from each language construct, and the frequencies of the machine instructions were estimated.

The SWARD instructions and their estimated frequencies (P_i) are shown in Table 17.8. The frequencies were adjusted slightly to account for machine instructions that do not directly correspond to language constructs. Inaccuracies in the estimated frequencies are tolerable, since only the relative rank of each instruction is needed for optimization. In fact, as seen later, minor inaccuracies in the ranking of instructions have a zero or negligible effect on the results.

Table 17.7 B1700 SDL Op-code Encoding Comparison

Method	Operating System Op-code Bits	Percentage Improvement	Decoding Penalty (%)
8-bit field	301,248	0	0
4-6-10	184,966	39	2.6
Huffman	172,346	43	17.2

The first step was an **H** calculation on the instruction frequencies. Treating the instructions as independent (which is reasonable, since "less primitive" instruction sets should have smaller serial dependencies among instructions), **H** is found to be 4.1 bits. Hence a fixed-size 2-token (8-bit) op-code would occupy twice as much space as is theoretically necessary.

Although the eventual encoding scheme has to represent op-codes in units of 4 bits (because this is the smallest unit of storage allocation), a Huffman encoding was done to establish the best possible actual representation. The op-code lengths using Huffman encoding are shown in Table 17.8. The average op-code length is 4.2 bits. These 4-bit-unit encoding methods were then considered: the "8/64/512" method and the "15/15/15/..." method.

In the 8/64/512 method, the eight most frequent instructions have a 1-token (4-bit) op-code, the 64 next most frequent instructions have a 2-token op-code, and so on. This could be accomplished in the following way. If the first bit of the first token is 1, the op-code is 1 token in length (1XXX). If the first bit of the first token is 0 and the first bit of the second token is 1, the op-code is 2 tokens (0XXX1XXX), and so on.

In the 15/15/... method, the 15 most frequent instructions have a 1-token op-code, the 15 next most frequent instructions have a 2-token op-code, and so on. If the first token is not zero, the op-code is represented as 1 token. If the first token is zero and the second token is not, the op-code is represented as 2 tokens, and so on.

Since it is not obvious which is better, the op-code lengths for both methods were determined as shown in Table 17.8. Multiplying each op-code length by the corresponding frequency and then summing the results yields an average op-code length of 5.06 bits for the 8/64 method and 4.63 bits for the 15/15/... method. Hence the latter method was selected.

ADDRESS OPTIMIZATION

As mentioned earlier, the same process can be used for other types of information such as data addresses, instruction addresses, and data. Here we consider removing redundancy from addresses.

Table 17.8 SWARD Op-code Encodings

Inst.	P_i	Rank	Huffman Bits	8/64 Bits	15/15/ Bits
MOVE	20.0	1	2	4	4
ADD	12.5	2	3	4	4
BF	11.0	3	3	4	4
B	9.0	4	4	4	4
CONVERT	9.0	5	4	4	4
LE	4.5	6	4	4	4
SUB	4.0	7	5	4	4
EQ	3.5	8	5	4	4
MOVESS	3.0	9	5	8	4
CALL	2.5	10	5	8	4
CONCAT	2.0	11	6	8	4
MULT	1.8	12	6	8	4
DIVIDE	1.8	13	6	8	4
LT	1.5	14	6	8	4
GT	1.5	15	6	8	4
COMP	1.0	16	7	8	8
POWER	1.0	17	7	8	8
NE	1.0	18	7	8	8
AND	1.0	19	7	8	8
LENGTH	0.7	20	7	8	8
OR	0.7	21	7	8	8
INDEX	0.6	22	8	8	8
ACT	0.6	23	8	8	8
RETURN	0.6	24	8	8	8
LCALL	0.6	25	8	8	8
GE	0.5	26	8	8	8
LACT	0.3	27	8	8	8
LRETURN	0.3	28	8	8	8
CPTR	0.3	29	8	8	8
ALLOC	0.3	30	8	8	8
FREE	0.3	31	8	8	12
CHAIN	0.3	32	9	8	12
UNDEF	0.2	33	9	8	12
MODULO	0.2	34	9	8	12
ABS	0.2	35	9	8	12
DEF	0.2	36	9	8	12
NOT	0.2	37	10	8	12
SIGNAL	0.2	38	10	8	12
CACC	0.1	39	10	8	12
LMODULE	0.1	40	10	8	12
LINK	0.1	41	10	8	12
ENABLE	0.1	42	10	8	12

Table 17.8 *(Continued)*

Inst.	P$_i$	Rank	Huffman Bits	8/64 Bits	15/15/ Bits
DISABLE	0.1	43	10	8	12
CONT	0.1	44	10	8	12
TRFAULT	0.1	45	10	8	12
DTAG	0.1	46	10	8	16
DCON	0.1	47	10	8	16
TRACE	0.1	48	10	8	16
NOTRACE	0.1	49	10	8	16

In addition to optimizing op-code representations in the SDL-oriented architecture of the Burroughs B1700, data addresses were also optimized [5]. Lexical-level addressing is used in this architecture. The designers decided to allow for 16 lexical levels and 1024 variables in each level, leading to a 14-bit lexical level address, 4 bits for the level number and 10 bits for the order or index number. After the compilers and the operating system were written in SDL, the actual addresses were analyzed. As would be expected, the values were significantly skewed, implying a significant amount of redundancy in the single-size representation.

As a result, the architecture was changed to frequency-base encode the addresses. Addresses can be 8, 11, 13, or 16 bits in length. An address contains three fields: a prefix, lexical level, and order number. The 2-bit prefix defines the format (size) of the address. The lexical-level field can be 1 or 4 bits in length. The order number is represented in 5 or 10 bits. A study of data addresses in instructions in the operating system showed that the 9174 addresses required 128,436 bits in the original version, but only 94,900 bits in the new version, a reduction of 26% in memory requirements.

Program addresses were optimized in a similar way. A program address consists of a segment number and a bit displacement of an instruction within the segment. The designers decided to allow for 1024 segments of one million bits in length, leading to a 30-bit instruction address (10-bit segment name and 20-bit displacement). However, some encoding was done in the original version, allowing 5- or 10-bit segment names and 12-, 16-, or 20-bit displacements. Also, the designers recognized that many references to instructions are intrasegment references (branches); thus a null (zero-bit) segment name was permitted. Furthermore, many references are to the beginning of another segment; thus a null displacement was permitted. Hence the program address has three fields: a 3-bit prefix defining the representation of the address, a 0-, 5-, or 10-bit segment name, and a 0-, 12-, 16-, or 20-bit displacement (only eight of the segment/displacement combinations are valid).

The distribution of program-address values in compiled programs was studied to judge the wisdom of this intuitive scheme. It proved to be close enough to optimal to warrant no changes. A study of 3767 program addresses in instructions in the operating system showed that 74,303 bits were needed, versus the 120,544 bits that would be needed in a byte-oriented machine with fixed-size addresses and the same addressing range, a saving of 38%.

Address Optimization in the SWARD Machine

The SWARD architecture is another illustration of the reduction of redundancy in addresses, although this was accomplished in a different manner than in the B1700. Address optimization results from three attributes of the architecture: (1) each module has its own address space, (2) data address spaces are distinct from instruction spaces, and (3) a module defines the sizes of its single-size data and instruction addresses.

Although only the third attribute has address optimization as its primary motivation, all three contribute to small addresses. Consider a machine in which all programs are represented in a single large address space or an entire program is represented in a single address space (i.e., most conventional machines). This implies that any address can potentially reference any storage location. However, because of constraints such as the semantics of the programming language, there is a highly skewed relationship between instructions and addresses. That is, the machine instructions corresponding to a subroutine in the source-language program reference only a small subset of all possible storage locations. Hence there is considerable redundancy in addresses. In the SWARD machine, much of this redundancy is removed; since each module has its own address space, data addresses are smaller.

Likewise, an architecture that places instructions and data in the same address space has unnecessary redundancy in its addresses. For instance, all storage locations are not equiprobable targets of branching instructions; storage locations holding data should have a probability of zero. Similarly, all storage locations are not equiprobable targets of an add instruction; in particular, storage locations holding instructions should never be referenced by an add instruction. Hence placing instructions and data in separate address spaces removes some of this redundancy and leads to shorter addresses.

The astute reader may have spotted two other sources of redundancy. The first is the observation that every variable in a program is not likely to be referenced an equal number of times. Hence address values are not equiprobable, and the **H** calculation shows that, when this is the case, redundancy exists if the values have the same size. The second, and more serious, source is that certain address values are never used because they do not represent the starting point of data representations (i.e., cells). For instance, consider a module in the SWARD machine with four cells. The length of the first cell is 4 tokens, the second is 6, the third is 9, and the last is 4. Hence the only valid cell addresses are 1, 5, 11, and 20. However, the other values (2, 3, 4, 6, etc.) can be represented in a cell address; therefore a considerable amount of redundancy must be present.

This type of redundancy was studied during the development of the SWARD architecture [6]. Two methods of addressing were analyzed: the "direct method" and the "indirect method." The direct method is the one with this apparent redundancy: an address field contains the offset of a cell within the module's address space. In the indirect method, an address contains the sequential number of a cell. The module would contain a table (called the cell-offset table); each entry (one per cell) contains the offset of the corresponding cell within the address space.

Since it is not obvious which method is better, they are compared from four points of view: (1) the physical size of the module, (2) the memory-processor data transfer to execute the module, (3) the amount of computation performed by the processor to translate an address field to a physical cell address, and (4) the locality of reference (e.g., if the processor has to fetch 2 tokens, it is more desirable for the tokens to be in contiguous locations). These identifiers are used in the analysis:

A number of address fields in the module. Address fields are found in instructions and in structure and relocatable cells.
F number of operands referenced during execution of the module.
C number of cells in the module.
S size (in tokens) of the module's address space.

The function CEIL is used to represent raising a number to its next higher integer (e.g., CEIL(2)=2, CEIL(2.1)=3). All logarithms are to the base 16.

ADDRESS OPTIMIZATION

Considering first the size of a module, the size will be the same except for the space occupied by addresses and the cell-offset table. Hence the constant factor is not considered. In the indirect method, the size is given by

$A \times \text{CEIL}(\log C) + C \times \text{CEIL}(\log S)$

which is the space occupied by the address fields plus the space occupied by the cell-offset table. In the direct method, the size is

$A \times \text{CEIL}(\log S)$.

The ratio of these formulae is

$((\text{CEIL}(\log C))/(\text{CEIL}(\log S))) + C/A$

If this expression is less than 1, the indirect method is better.

The only relationship that is always true is $S > C$; thus one method cannot be determined to be universally better. However, it is likely that S is much greater than C (an order of magnitude seems reasonable); thus the first component will typically have the values 1.0, 0.67, or 0.5. If we assume that, for all modules, C is uniformly distributed over the range of 2–300 and that $S = 10 \times C$, the expected value of the first component is 0.72. The relationship $A > C$ should be true, for if it is not, the module would contain variables that are not elements of structures and that are never referenced. If we assume that a typical module has an average of three references to each variable, the ratio will be greater than 1. For instance, the ratio is 1.24 for the program in Figures 14.12 and 14.13. Hence the direct method seems to be more efficient in terms of space.

Considering the memory/processor data transfer, both methods show a constant factor that can be omitted (transmission of operation codes, literals, and cells); thus the analysis can be limited to the amount of addressing information transmitted. For the indirect method this information is

$F \times \text{CEIL}(\log C) + F \times \text{CEIL}(\log S)$

which represents a cell number and cell offset for each cell reference. For the direct method the information is

$F \times \text{CEIL}(\log S)$

which clearly indicates that the direct method is better.

Considering address computation, the indirect method involves more computation by the processor, since the cell number must be multiplied by the width of the entries in the cell-offset table to obtain the offset value. Hence the direct method is again superior.

Data-transmission calculations do not provide a totally accurate comparison, because a method with a larger amount of transmitted data could result in fewer memory fetches if the data were contiguous and if the data fetched by the other method were discontiguous. Hence the locality of reference should be considered. However, this consideration does not aid the indirect method; the indirect method requires the transmission of more data, and these additional data are discontiguous (i.e., a cell-offset-table entry is discontiguous from an address field).

Because of this analysis, the direct method (addressing by cell offset) was chosen over the indirect method (addressing by cell number).

Although this discussion about making time and space improvements by reducing redundancy at the hardware/software occurs toward the end of this book, the reader should now consider that the entire book is about redundancy removal. For instance, we have seen that a language-directed architecture enables one to express a program in fewer bits than in a conventional von Neumann architecture. Hence a language-directed architecture is a redundancy remover. The concept of self-identifying storage (tagged storage) is also a redundancy remover; in such an architecture, the attributes of a datum are stored with the datum rather than repeating the attributes in each machine instruction that references the datum.

This chapter merely scratches the surface of specific optimization techniques. For instance, another well-known phenomenon is that the targets of most branch instructions are likely to be only a short distance away, implying that, rather than specifying branch addresses relative to some fixed location, they should be specified as relative to the current value of the instruction counter. Hehner discusses the optimization of operation codes, addresses, and data in more detail [8–10]. One of his studies was an analysis of a 2137-statement XPL program (XPL is a language that resembles PL/I). When compiled to the IBM S/360, the program occupied 486,704 bits. When compiled to a hypothetical language-directed machine, the program occupied 299,696 bits, 62% of the S/360 version. After extensive optimizations were performed on the hypothetical machine, the program occupied only 118,063 bits, or 24% of the S/360 version and 39% of the version on the unoptimized hypothetical machine.

EXERCISES

REFERENCES

1. W. G. Alexander and D. B. Wortman, "Static and Dynamic Characteristics of XPL Programs," *Computer*, **8**(11), 41–46 (1975).
2. L. Svobodova, "Computer System Performance Measurement: Instruction Set Processor Level and Microcode Level," SEL-74-015, Digital Systems Laboratory, Stanford University, 1974.
3. D. B. Wortman, "A Study of Language Directed Computer Design," Ph.D. dissertation, Stanford University, 1972.
4. R. Ash, *Information Theory*. New York: Wiley, 1965.
5. W. T. Wilner, "Burroughs B1700 Memory Utilization," *Proceedings of the 1972 Fall Joint Computer Conference*. Montvale, N.J.: AFIPS, 1972, pp. 579–586.
6. G. J. Myers, "The Design of Computer Architectures to Enhance Software Reliability," Ph.D. dissertation, Polytechnic Institute of New York, 1977.
7. J. L. Elshoff, "An Analysis of some Commercial PL/I Programs," *IEEE Transactions on Software Engineering*, **SE-2**(2), 113–120 (1976).
8. E. C. R. Hehner, "Matching Program and Data Representations to a Computing Environment," Ph.D. dissertation, University of Toronto, 1974.
9. E. C. R. Hehner, "Computer Design to Minimize Memory Requirements," *Computer*, **9**(8), 65–70 (1976).
10. E. C. R. Hehner, "Information Content of Programs and Operation Encoding," *Journal of the ACM*, **24**(2), 290–297 (1977).

EXERCISES

17.1 In the optimized version of the Student-PL machine, what is the meaning of these instructions?

   (a) 10100100
   (b) 00100011 (in lexical-level 1)
   (c) 00000011 (in lexical-level 1)
   (d) 00100011 (in lexical-level 3)
   (e) 00000011 (in lexical-level 3)
   (f) 01100011 (in lexical-level 2)

17.2 From the discussion of the SNAME, SLOAD, and LITERAL instructions for the Student-PL machine, what can you deduce about the values of the op-codes for the other instructions?

17.3 Compile the second Student-PL program in Chapter 6 to the optimized Student-PL machine. Compare your result in terms of space with the program in Figure 6.4.

17.4 Would you expect to find greater or fewer opportunities for instruction-pairing optimization in a language-directed machine than in a conventional machine?

17.5 What is the actual average amount of information in an op-code for a five-instruction machine where the probabilities of occurrence of the instructions are 0.5, 0.3, 0.1, 0.05, and 0.05, assuming no serial dependencies among the instructions? What is the redundancy if the machine has a fixed-size 3-bit op-code field? How does the actual average amount of information change if serial dependencies exist (i.e., given one instruction, the probability of the following instruction is different from the probabilities listed above)?

17.6 Although only two op-code encoding schemes were compared for the SWARD machine (8 /64 /512 and 15 /15 /. . .), a large number of other alternatives are possible. Design an op-code representation such that 14 instructions have a 1-token (4-bit) op-code, 28 instructions have a 2-token op-code, and 56 instructions have a 3-token op-code.

17.7 Determine if the encoding in the previous exercise is better than the 15 /15 /. . . method.

17.8 Under what conditions, in a machine with a fixed-size 8-bit op-code, would **H**=8?

17.9 In the 4-6-10 op-code encoding in the SDL architecture in the Burroughs B1700, how many 4-, 6-, and 10-bit op-codes can be defined?

18 | The Art of Computer Architecture

The focus of attention of this book is primarily a set of architectural concepts that reduce the troublesome traditional semantic gap between programming languages and computer architectures, using several actual or proposed machines to illustrate these concepts. However, little has been said about the methodologies or thought processes used to produce a computer architecture. It seems only natural, then, to conclude with a short discussion of a few goals and tools that should be considered by the computer architect.

CONCEPTUAL INTEGRITY

The term "conceptual integrity" was coined as an objective for designing the external interfaces of a software system [1], although the concept is applicable to computer architecture as well as other disciplines. Conceptual integrity, or "integrity of concept," is synonymous with the terms *uniformity, consistency,* and *regularity*. When applied to computer architecture, it implies that the architecture should exhibit a complete and consistent structure, that special or exceptional cases should be absent or severely minimized, and that the architecture should present the impression of having been conceived, designed, and documented by a single mind.

Conceptual integrity in a computer architecture is important for two reasons: ease of programming (or code generation) and efficiency. It is not difficult to see that a lack of consistency in an architecture, or a large number of special cases, leads to increased programming time, more mistakes, and increased debugging difficulty in assembly-language programs. A lack of conceptual integrity can tremendously complicate the code-generation portion of a compiler. More important, a large number of special cases in an architecture can lead to decreased program efficiency because of the difficulties in dealing with the special cases in the compilers' optimization phases.

Architectural concepts such as self-identifying data with an associated generic instruction set and the one-level store are examples of increases in conceptual integrity, but the best way to understand the idea is to look at counterexamples. One can find counterexamples in almost any architecture. To illustrate what conceptual integrity is not, a few examples from the IBM S/370 are used. The S/370 is not used because it has less conceptual integrity than other architectures, but because most readers are probably familiar with its architecture. A few examples of irregularity or special cases in this architecture are

1. The machine contains instructions to convert binary numbers to decimal and vice versa, but no conversion instructions exist for other data representations (e.g., floating point).
2. For many instructions that require the use of multiple general-purpose registers, two consecutively numbered registers must be used. The instruction must name the lower-numbered register, and this register must have an *even* number.
3. If an overflow occurs in a convert-to-binary instruction, the exception generated is "fixed-point divide."
4. Decimal (base 10) numbers can only have an odd number of digits of precision.
5. Two binary number representations are present: 32 bits (fullword) and 16 bits (halfword). Most of the fullword-oriented instructions (e.g., load, store, add, subtract, multiply) have corresponding halfword-oriented instructions, with the exception of the divide instruction.
6. A few instructions (e.g., move-character-long) require the use of four general-purpose registers, complicating the optimization of register assignments.
7. The general-purpose registers are almost, but not quite, "general purpose." One exception is the translate-and-test instruction, which always uses registers 1 and 2.
8. There is a halve (divide by two) instruction for floating-point numbers, but none for binary or decimal numbers.

TECHNOLOGY INDEPENDENCE

ORTHOGONALITY

When read literally, orthogonality implies a "right-angled" relationship among concepts in the architecture. It is the objective of (1) holding the number of basic concepts to a reasonable minimum, (2) maximizing the independence among the concepts, and (3) avoiding superfluities. In other words, an architecture with low orthogonality has a large number of overlapping concepts, a number of "nice looking" but not particularly useful operations, and 17 different ways to zero or increment a register. An architecture with high orthogonality tends to provide more function at a given level of complexity or cost.

EXTENSIBILITY

The ability to extend the architecture in a compatible fashion is another important consideration, although it is difficult to discuss this objective in a general way. Among other things, extensibility implies that arbitrary restrictions on the number of instruction and data types and on the maximum addressing range should be avoided. Some of the time and space optimizations discussed in Chapter 17 (e.g., frequency-based encoding of op-codes, addresses, and data) contribute to extensibility.

IMPLEMENTATION FREEDOM

Ideally, an architecture should place few, if any, constraints on the implementation of the underlying processor. For instance, the architecture should allow one to create a wide range of underlying implementations with different cost and performance levels (i.e., a family of compatible machines). The "less primitive" instruction sets of non-von Neumann machines allow more implementation freedom than those of a von Neumann machine. The instructions tend to contain a significant amount of opportunity for parallelism, meaning that the instructions can be implemented sequentially in a slower and cheaper implementation, yet there is a significant opportunity for creativity in implementing a large-scale processor for such an architecture.

TECHNOLOGY INDEPENDENCE

Since logic and memory technologies are advancing rather quickly, another objective is avoiding any constraints on the exploitation of technologies used to construct the underlying system. For instance, capability-based addressing and the one-level store hide, at the architecture level, the amount, types, and characteristics of the physical storage media.

FORMAL DESCRIPTION

Another consideration during the development of an architecture is the use of a formal description language for the specification (or as a supplement to a prose specification). Three main benefits of this are (1) the formal description is likely to be more precise and less ambiguous than a prose specification, (2) if the formal description is a semantic or algorithmic definition of the architecture, it is useful as a guide when designing a processor to embody the architecture, and (3) if the description language is executable, it can serve as a model of the architecture for performance studies and the testing of actual implementations.

The most widely known description language is ISP [2]. Unfortunately, ISP is oriented largely toward expressing the "syntax" of an architecture (e.g., instruction and data formats); it is weak as a tool for expressing the semantics of an architecture. As a few other examples, the Student-PL machine was defined with a procedural language called MDL [3], the IBM S/360 has been defined in APL [4], and the SWARD machine was defined in a slightly extended dialect of PL/I [5]. The use of a formal description language for the SWARD machine points out a fourth advantage: 13 problems were discovered (and corrected) in the earlier prose description during the act of writing the formal description.

MENTAL COMPILATION

Throughout the book, the practice of mentally compiling a high-level-language program to an architecture has been encouraged as a vehicle for understanding the architecture. In fact, this practice is an effective technique during the development of an architecture to analyze alternatives and discover problems. As an example, the PL/I programs in Figures 14.9 and 14.10 were repeatedly compiled, during the development of the SWARD architecture, to each interim version of the architecture.

LANGUAGE VALIDATION

Although it is not absolutely true, it is usually true that, given that a compiler can be written to compile programming language L to a von Neumann machine A, L can also be compiled to a different von

Neumann machine B. That is to say, when designing a von Neumann architecture, one (hopefully) may worry about the efficiency of compiled programs, but one rarely, if ever, worries about the impossibility of compiling certain languages to the architecture.

However, in a non-von Neumann architecture, the latter issue—whether it is even *possible* to compile a particular language to the architecture—is an important consideration, although, to the author's knowledge, this issue has never been posed in the literature. In other words, because non-von Neumann architectures remove much of the dangerous generality inherent in von Neumann architectures and because current languages were designed with von Neumann architectures in mind, it may be impossible to compile certain language constructs to the architecture.

Language validation, then, is the process of determining the feasibility of compiling a particular language to an architecture. Since this issue has never been posed, no methodologies for this process have been developed. However, an informal methodology that might be used is to "prove" that each language construct (e.g., data type, statement, operator) can be mapped to one or more constructs in the architecture. For instance, if one studied the relationship between PL/I and the Student-PL machine, it would be seen that writing a PL/I compiler is impossible (e.g., the important concept of a PL/I structure cannot be represented). If one studied the relationship between FORTRAN and the SWARD machine, it would be seen that certain types of EQUIVALENCE statements (e.g., "equivalencing" an integer variable and a real variable) cannot be represented. (This was recognized during the development of the SWARD architecture; it was not provided for because it is considered a poor programming practice—it makes the program machine dependent and is a source of programming errors.)

REFERENCES

1. F. P. Brooks, *The Mythical Man-Month: Essays on Software Engineering.* Reading, Mass.: Addison-Wesley, 1975.
2. C. G. Bell and A. Newell, *Computer Structures: Readings and Examples.* New York: McGraw-Hill, 1971.
3. D. B. Wortman, "A Study of Language Directed Computer Design," Ph.D. dissertation, Stanford University, 1972.
4. A. D. Falkoff, K. E. Iverson, and E. H. Sussenguth, "A Formal Description of System/360," *IBM Systems Journal*, **3**(2), 198–261 (1964).
5. G. J. Myers, "The Design of Computer Architectures to Enhance Software Reliability," Ph.D. dissertation, Polytechnic Institute of New York, 1977.

Answers to Exercises

1.1 Yes and no. Yes, because most existing systems were defined this way, although the end result is usually far from optimal. If the question began with the word "Should," the answer would be no, since it is inconceivable that someone could intelligently distribute the system's functions among software and hardware unless he or she has first defined the system's functions and the languages to be provided.

1.2 The compiler designer. Ideally, the application programmer should know nothing about computer architecture. The operating system designer is likely to have the wrong bias, because, although operating system efficiency is important, the objective of most systems is to execute the end-use applications as efficiently as possible. The processor engineer is probably the worst candidate unless he or she has significant experience above the interface (i.e., software experience).

2.1 Reducing the semantic gap too far may actually increase it. Consider designing a machine such that its semantic gap to PL/I is reduced almost to zero. This may be fine for PL/I programs, but one is likely to find that the semantic gap between this architecture and other languages (e.g., COBOL, RPG) is now greater than the gap would have been between these languages and a conventional architecture. Hence one strategy for a multilanguage system is to reduce the gap between the architecture and the semantic aspects that the languages have in common, without orienting the architecture too far in the direction of one language. There are other strategies; these are discussed in the architectures of Parts IV and V.

2.2 The related characteristics are (1) rather than a single sequential address space, storage should more closely resemble the concept of storage in programming languages, (2) storage should have the ability to be viewed as multidimensional, (3) the machine should know the distinction between data and programs, and (4) data in storage should have a self-identifying property.

2.3 If the numbers are represented in base two, their base two values are only approximations of their base 10 values (e.g., 0.1 in decimal is equivalent to the endless 0.0001100110011 ... in base two, which must be truncated to a finite length).

2.4 20.4%. The number of bits occupied by decimal-encoded numbers relative to their binary equivalents is $4 \log 2$ (assuming all number values are equiprobable).

2.5 Although it is unwieldy, one could represent each pair of decimal digits in 7 bits (base 100 arithmetic). The inefficiency is reduced to 5.4% $(1-(7/2)\log 2)$.

3.1 Five possible solutions are
a. The problem may be of no significance if the system is intended to be used with only a single language.
b. In the language-directed approach, one could direct the architecture toward the common semantics of programming languages rather than one particular language.
c. One could configure the system into multiple processors, each associated with a particular high-level language, and have the operating system schedule programs to the appropriate processor.
d. Similarly, the processor could have multiple architectures (e.g., multiple microprograms). The operating system would switch the processor to the appropriate microprogram when scheduling a program (the Burroughs B1700 approach).
e. One could provide a basic instruction set usable by all languages and multiple supplemental instruction sets oriented toward particular languages (the approach in the SWARD machine discussed in Part V).

3.2 Both. If the system supports high-level programming languages, the semantic-gap problem exists independent of the scale of the system. As evidence, the references in Chapter 3 cover the spectrum from large-scale systems to small minicomputers for aerospace applications.

3.3 The two most widely used languages, COBOL and RPG, have received almost no attention.

4.1 Overlaying or aliasing of storage, such as the EQUIVALENCE statement in FORTRAN, the REDEFINES clause in COBOL, or the DEFINED attribute in PL/I, if the storage area is defined as representing variables of different attributes, such as

    INTEGER A
    REAL B
    EQUIVALENCE (A,B)

The only solace is that this is considered by many to be a poor programming practice.

4.3 A program on machine Y requires less space if R is greater than 1.6.

4.4 A program on the tagged machine (Y) occupies 85% of the space occupied by the same program on machine X. Machine X has 48-bit instructions; Y has 38-bit instructions. The average operand size on X is 32 bits; on Y it is 40 bits. The assumptions imply a ratio of five instructions per operand.

4.5 No, since FORTRAN statements do not operate on entire arrays. It is probably asking too much of a compiler to recognize that the loop

    DIMENSION A(10),B(10)
    DO 20 I = 1,10
      A(I) = A(I) + B(I)
    20 CONTINUE

could be translated into a single generic machine instruction (add B to A). This is an example of going overboard by providing concepts in the architecture that are unusable by the compiler. (However, if the system also contained a PL/I compiler, the answer might be the opposite.)

4.6   PUSHAD A
      PUSH   B
      PUSH   B
      PUSH   B
      MULT
      ADD
      STORE

4.7 Five, three, and one, respectively.

4.8 Eight, nine, and three, respectively.

4.9 One hundred, 96, and 48, respectively.

4.10 In terms of the number of instructions generated, the number of units of information to be decoded, and the size of the instruction stream, the register and stack forms appear to be roughly equivalent, but the storage-to-storage form appears to be superior.

5.1 The type field defines the data type of the array elements.

5.2 The string segment, symbol table, and scope table memories.

5.3 The symbol table and scope table memories.

5.4 An array pointer in the data stack points to a descriptor in the array descriptor memory. The descriptor in the array descriptor memory points to the first array element in the upper portion of the data stack memory. The array elements in the data stack memory would be character-string descriptors pointing to the strings in the string segment memory.

5.5 Sixteen (because there are 16 display registers).

5.6 The data stack memory and the DSP register (for the allocation of local storage for the called procedure), the string segment memory (if the called procedure contains character variables), the program segment stack and scope stack memories and their PSP and SSP registers, and the display registers.

5.7 Not always. Each entry would point to the prior stack entry, except in the case of recursive invocations of procedures.

6.1 DS(103) = array pointer, DS(102) = the value 0, DS(101) = the value 1.

6.2 Their purpose is initializing the first array element (subscript 0) as a program-segment descriptor of the segment at location 700 (the ELSE alternative).

6.3 The program-segment descriptor on the stack points to the segment being entered. The first instruction of the segment must be SCOPEID, which points to the scope table entry.

6.4 The number of arguments is specified by the instruction's operand. The number of parameters is specified in the scope table entry.

6.5 The first one removes the undefined value returned from ZZZ. The second one removes the argument indirect-address descriptor placed on the stack by instruction 34.

6.6 The scope table entry specifies the number of parameters, the

ANSWERS TO EXERCISES 303

location of the first symbol table entry, and the number of entries. The parameters always appear as the first symbol table entries, because they are the first names encountered in compiling a procedure (i.e., they are listed in the PROCEDURE statement).

8.1 The obvious shortcoming is that the system is costly to change. For instance, the addition of a new statement to SPL would require a partial redesign of the sequential logic network in the translator.

8.2 Access the fourth element of the third element of the second element in structure X.

8.3 It could be either. An element of a structure can be a scalar or a structure.

8.4 A: I-200-300-WXYZ-null-I
    101   100
     I   I
    102   200
                           I
                        300
                         I
                      WXYZ

9.1 The numeric internal form is a packed-decimal representation of a numerical value. Rather than storing each digit in 8 bits, two decimal digits are stored per 8 bits.

9.2 FD XX XX XX XX XX XX
    F5 30 30 30 30 30 30
    F6 XX XX XX XX XX F6
    F5 31 31 31 31 31 31
    F6 XX XX XX XX XX F6
    FF XX XX XX XX XX XX

9.3 F5 31 F6 XX XX XX XX F6

9.4 80 YY YY YY XX XX XX XX
where YYYYYY is the memory address of the area that was dynamically allocated to hold the string 1234.5678.

9.5 They are necessary because the BLOCK instruction is the target of a GO TO (it corresponds to label L). If the BLOCK instruction were moved up to word 3007, the LOAD-SOURCE-POINTER instruction in 3007 would execute after each GO TO, but this instruction is associated with the previous (3004-3006) object code. The program would work properly with both no-ops removed, but the programmer might be directed to the wrong source statement if an error occurred.

ANSWERS TO EXERCISES

9.6 The identifier control word at location 2222. The identifier control word at location 2272.

10.1 The terminal/user identification must be transmitted to tell the memory controller to which user's page lists the request is associated (i.e., for an assign-group request).

10.2 The central processor uses a logical-storage string to represent its stack. It uses fetch-and-reverse-follow as a "pop" operation.

10.3 If the IGAC were not present, the memory controller would have to link all the group-link words together to form the available-group list whenever a free page was placed into use. With the IGAC, the memory controller accomplishes the same purpose by simply storing the value 1 in the IGAC when placing a free page into use. If the page is used for a long period of time and its groups are allocated and reclaimed, the IGAC will eventually be incremented to 29; groups are then taken from the available-group list. Hence the IGAC reduces the overhead of the page-allocation process.

10.4 To efficiently implement the fetch-and-reverse-follow operation.

10.5 The operations are provided because the memory controller and memory reclaimer are manipulating the same lists. If the memory reclaimer were in the process of removing something from a garbage list or placing something on an available list and, in the midst of this, the memory controller used these lists on behalf of another processor, timing errors could easily occur (e.g., storage might be lost or placed on multiple lists). Therefore these operations are provided by the memory controller to serialize the manipulation of garbage and available lists.

10.6 No. This situation occurs only for strings representing SPL structures; these strings are always allocated from TPL2. Therefore the memory reclaimer need do this only if it sees that the page containing the string is on TPL2. The reference processor also assists the memory reclaimer by setting bit 37 in the group-link word (Table 10.3) if an address is being stored into one of the group's words.

10.7 Addresses and data values are self-identifying. Data values start with the control character F0, F1, F2, F3, or F5; substructure pointers start with the control character EC.

10.8 There is one available-page list in the system; it contains all the unallocated pages. There are three available-space lists per user; they contain the pages containing some available groups. There is one available-group list per page; it contains all the available reclaimed groups on the page.

ANSWERS TO EXERCISES

10.9 A[2,2] currently has the scalar value 202. A three-word string would be allocated to contain the vector <1234/4321>, and a substructure pointer to this string would be placed in the word currently containing the value 202. Also, the second word (a substructure pointer) in the leftmost box of Figure 10.4 would now contain a 02 subscript, and its last field would point to this new substructure pointer.

12.1 Six.

12.2 Sixteen bits (3 for the opcode, 7 for the first field, and 6 for the second field).

12.3 Nineteen bits.

12.4 Move the unsigned 4-bit data type of value 73 into the storage associated with the descriptor at descriptor table entry 27.

12.5 In hexadecimal, 4EF7F340.

12.6 $****10.45 (in EBCDIC).

12.7 The most striking attributes are that the majority of the instruction set is associated with editing and comparison operations. This represents the nature of most COBOL applications; COBOL programs tend to be highly oriented toward input/output and data-manipulation operations and perform relatively little arithmetic.

14.1 The cell represents a floating-point number whose value is -0.7493×10^{23}.

14.2 The output of the compilers, an object module, is represented in a token string. A module is defined to the machine by executing a LOAD-MODULE instruction; one of the operands of this instruction is a token string representing a module.

14.3 Three cells are represented. The first is a boolean string of size 3 and current length 2. Its value is false,true. The second cell is a 3-digit decimal integer. Its value is -43. The third cell is a two-dimensional 10×10 array of pointers. The array currently has no storage allocated for its elements.

14.4 10F00F9E

14.5 100F000109E.

14.6 10F6B9E.

14.7 A global variable would be represented as a relocatable cell, relocatable from a pointer cell. The linkage editor would initialize the pointer cell using the LINK instruction. In the case of a FORTRAN COMMON area, or a PL/I structure with the external attribute, the relocatable cell would be a structure.

14.8 One situation is that in which the language specifies that the value of the iteration variable of a DO loop is undefined when the loop terminates.

14.9

```
    Indexes                                        Comments

    001       000320003A0003E0003E                 Header
    015       0003E0009E0009E00114                 Header
    029       220000000                            Header
    032       E2D6D9E3                             Module name
    03A       0001                                 Entry-point list
    03E   01  671D8FFFFF00000000000000000000  X(N)
          1F  D80000F0000000                       SAVE
          2D  6FAF0000000000000000000000      N
          45  FA0F000000000                        I
          52  FA0F000000000                        J
          5F  CF                                   boolean ($B)
    09E   01  0902012D                        ACT  2,X,N
              85F2D002                        LT   $B,N,'2'
              F5F18                           BF   $B,%A
              0A                              RETURN
          18  145002                    %A:   MOVE I,2
          1E  152001                    %G:   MOVE J,1
          24  045FQ1450152              %E:   GE   $B,X(I),X(J)
              F5F38                           BF   $B,%B
              E4F                             B    %C
          38  11F0145                   %B:   MOVE SAVE,X(I)
              101450152                       MOVE X(I),X(J)
              101521F                         MOVE X(J),SAVE
          4F  85F5245                   %C:   LT   $B,J,I
              F5F64                           BF   $B,%D
              352001                          ADD  J,'1'
              E24                             B    %E
          64  85F452D                   %D:   LT   $B,I,N
              F5F77                           BF   $B,%F
              345001                          ADD  I,'1'
              E1E                             B    %G
    112   77  0A                        %F:   RETURN
```

16.1 A solution is
MAX [ITEM(VOL)] [1]
MARK (A) [ITEM: ITEM.VOL=REGF1]
CROSS-MARK (A) [EMP: EMP.DEPT = ITEM.DEPT: ITEM.MKED (A)]
ADD [EMP(SAL): (EMP.MKED(A)) AND (EMP.SAL<20000)] [100]
RESET(A) [EMP]
RESET(A) [ITEM]
EOQ

ANSWERS TO EXERCISES

16.2 Approximately 350 milliseconds as a worst case. On the average there will be a 10 millisecond delay to get to the start of the tracks to begin the program. MAX takes slightly more than one revolution, but the MARK instruction cannot begin until the start of the tracks appears again; thus MAX contributes 40 milliseconds. MARK, ADD, and the two RESETs take 20 milliseconds each. Making a worst-case assumption that the eight employees reside on different tracks, the CROSS-MARK instruction takes 220 milliseconds.

16.3 Other performance increases stem from (1) a significant reduction in the amount of data transferred between the device and main storage (in a conventional system, the searching logic is in the central processor), (2) a reduction in the number of instructions executed in the central processor for the data-management software, and (3) no need to search and update access paths (e.g., indexes).

16.4 The other advantage is the space requirement. An associative store contains only data; there is no need to maintain access paths (e.g., indexes), which, in a conventional system, occupy a significant amount of space.

16.5 MOVE A,B

17.1 (a) LITERAL instruction. Place a FIXED integer with the value +36 on the top of the stack.
  (b) SNAME instruction. Place an indirect-address descriptor corresponding to lexical-level address 1,35 on the top of the stack.
  (c) SNAME instruction. Place an indirect-address descriptor corresponding to lexical-level address 1,3 on the top of the stack.
  (d) SNAME instruction. Place an indirect-address descriptor corresponding to lexical-level address 1,3 on the top of the stack.
  (e) SNAME instruction. Place an indirect-address descriptor corresponding to lexical-level address 3,3 on the top of the stack.
  (f) SLOAD instruction. Place on the top of the stack the value stored at lexical-level address 1,3.

17.2 The value of the first 2 bits of the other op-codes must be 11. If this were not true, the machine would not be able to distinguish unambiguously among instructions.

17.4 Probably fewer, but it is difficult to say. On one hand, the instruction sequences in the language-directed machine, since its instructions are less primitive, are more dependent on how the source program is written than on the sequences of instructions emitted by compilers to create a particular language construct. For instance, there are not many obvious opportunities for this type of optimization in the SWARD machine. On the other hand, language-directed machines tend to have fewer instructions (and therefore fewer unique pairs of instructions), raising the frequency of a particular instruction pair.

17.5 **H** = 1.79 bits. The redundancy is 40%. If serial dependencies exist, the **H** of the op-code stream is less than 1.79 bits.

17.6 Rather than having 0000 be the escape code, define 0000 and 1111 as the escape codes. For example, let XXXX be 4 bits, where the value is not 0000 or 1111. Four-bit op-codes have the form XXXX. Eight-bit op-codes have the form 0000XXXX or 1111XXXX. Twelve-bit op-codes have the form 00000000XXXX, 00001111XXXX, 11110000XXXX, and 11111111XXXX.

17.7 Yes, slightly better. The average op-code length would be 4.54 bits, compared to 4.63 bits.

17.8 It would occur if all instructions are equiprobable and if there are exactly 256 unique op-codes, an unlikely set of circumstances.

17.9 There can be ten 4-bit op-codes, twenty six-bit op-codes, and 64 ten-bit op-codes (or 63, if an escape code for additional op-codes is reserved).

Index

Access code, 54
Access path, 252
Activation record, 49–53, 185, 187, 193, 196, 200–201, 212, 244–246
Activation stack, 49–50, 51–52, 188
ADAM-language machine, 29
Addressing by content, 252
Addressing by location, 252
Address optimization, 277–278, 284–290, 295
Algol, 27, 28, 29, 30, 50
APL, 27, 28–30, 103, 296
Application-directed architecture, 26, 31–33
Argument transmission, 105, 121, 181, 209
Array, descriptor, see Descriptor
  PL/I, 11–12, 26
  specialized processor, 32
  subscript checking, 16–17, 45, 79, 161, 204, 213, 222
  subscript error, 14–15, 81, 181
  subscripting, 44, 79, 85, 112, 142, 144, 157, 202–203, 213
  tagged storage, 42, 187
  and von Neumann architecture, 20
Associative registers, 54
Associative storage, cost, 254
  definition, 252–254
  model, 252–254
  one-level store, 269
  performance, 272
  RAP, 255–267
  rotating, 254, 255, 256
  systems, 255

Asynchronous input/output, 10
Atlas system, 10
Authority code, 54
Automatic data conversion, B1700, 157, 163
  overhead, 42
  semantic gap, 13
  source of errors, 18
  Student-PL machine, 73, 85–88
  SWARD machine, 226–231
  SYMBOL system, 143
  with tagged storage, 38–39

Back-end processor, 5, 251
BALM language, 27
Based variable, 197, 216, 219
Base ten, see Decimal number representation
Base two, see Binary number representation
BASIC, 30, 152
Binary number representation, 13, 21–22, 182, 186
Bit strings, 18
BLM, 27
Block structure, 12–13, 27, 50, 103, 106
Burroughs B1700, application-directed architecture, 33
  Cobol S-language, 156–174
  comparison, 154–155
  data types, 157
  descriptor table, 159–162
  language-directed architecture, 28, 151
  op-code encoding, 161, 282–284, 292

309

INDEX

processor design, 153–154
program parameters, 157–158
SDL-oriented architecture, 153, 283, 286, 292
S-language, 153
subroutine management, 173
Burroughs B1800, 154
Burroughs B3500, 27, 154
Burroughs B5500, 27
Burroughs B6500, 27, 183
Burroughs B6700, 27, 43–44, 183
Burroughs B7600, 27
Burroughs Interpreter, 28

Cache memory, 9, 45, 154, 250, 268
Call /return mechanism, see Subroutine management
CAL operating system, 188
Capability-based addressing, definition, 52–55
  forerunner, 156, 180
  one-level store, 268–270
  SWARD machine, 185–186, 188, 193–194, 268–270
  technology independence, 296
Channel, 5, 10, 99, 100, 261, 266
Cobol, B1700, 152–153, 154–155
  language-directed architecture, 27–28, 156–174
  PERFORM statement, 27, 89
  REDEFINES, 223
  SWARD machine, 198, 199, 223
  variable-size data, 22
Compiler, as assembler, 28
  B1700, 152–153, 283
  code generation, 13, 16–17, 20–21, 23, 25, 294
  and computer architect, 7
  and conceptual integrity, 294
  and descriptor, 45
  detection of syntax errors, 31
  error detection, 181, 183
  heuristic instruction set, 32
  and register-oriented instruction set, 47–49
  and stack-oriented instruction set, 47–49
  SWARD machine, 198, 199, 242–244
  SYMBOL system, 98, 102, 144–145
  and tagged storage, 38, 40
Computer architect, 6–7

Computer architecture, 5, 6, 9
Computer-architecture methodologies, 293
Conceptual integrity, 293–294
Configuration architecture, 5, 251
Content-addressable storage, 252. See also Associative storage
Control Data Star-100, 251
Controlled storage, 13, 62, 66
Cost /performance ratio, 30–31

Dangling reference problem, 181, 186
Data base processor, 5, 251. See also RAP
Data base system, 40, 249–252, 256
Data independence, 40
Data scoping, see Scope of names
Datatron computer, 10
Data-value distribution, 22, 278
Debugging, 14–15, 40, 180, 183, 188, 210, 222
Decimal number representation, 13, 17–18, 21–22, 104, 156–157, 186, 191–192
Descriptor, B1700, 156, 159–162
  B6700, 43–44
  definition, 43–46
  Student-PL machine, 64–66, 71, 84–85
  SWARD machine, 186, 188
Device independence, 267, 270
Display registers, 52, 64, 67, 70–71, 73, 79
Distributed processing, 125
Domain, 257, 258–259
Domain name, 257, 258–260
Dynamic instruction frequencies, 274, 276–277, 280
Dynamic storage management, 110, 117

ECAM, 255
EDSAC, 10
EDVAC, 10
Efficiency, 7, 294
EULER machine, 29
Expression-evaluation stack, definition, 46–47
  evaluation, 48–49, 57
  Student-PL machine, 63–64, 69–73
  SWARD machine, 186
  SYMBOL system, 117
Extensibility, 295

File-management system, 249
File-storage station, 251

INDEX

Firmware, see Microprogram
Fixed-size storage words, 22–23, 48, 55, 182
Floating-point representation, 10, 13, 22
FLUID language, 28
Formal description, 296
Fortran, arithmetic IF, 18
  B1700, 152–153, 154
  COMMON area, 20, 224
  DO loops, 18
  EQUIVALENCE statement, 223, 297
  high-level-language architecture, 28–29
  language-directed architecture, 27–28
  language study, 48
  software errors, 184
  SWARD machine, 198, 223, 224
Free-variable operation, 257
Frequency-based encoding, 202, 295. See also Optimization
Front-end processor, 251

General-purpose registers, 10, 23, 294
Generic instructions, 38, 44, 157, 188, 268, 294
Global data, 122, 145, 181, 182, 224

H-calculation, 279–280, 288, 292
Heuristic instruction set, 32
Hewlett-Packard HP3000, 49
High-level-language machine, 26, 28–31
Honeywell Series 60, 32
Huffman encoding, 281–286

IBM 3838, 32
IBM 3850, 250
IBM 5100, 29–30
IBM 704, 10, 18
IBM 709, 10
IBM experimental architecture, 32
IBM Intermediate Language Machine, 27
IBM NORC, 10
IBM PL/I Checkout Compiler, 14
IBM PL/I Optimizing Compiler, 15, 16, 42, 56, 80, 183–184
IBM System /3, 30, 155
IBM S/360-370, addressing, 158, 181
  APL assist, 28
  architecture, 6
  array-processing attachment, 32
  binary arithmetic, 18, 294

comparisons to case studies, 80–81, 123, 154–155, 183–184, 215, 222, 279, 290
decimal arithmetic, 17, 294
formal description, 296
and IBM 5100, 30
instruction frequencies, 274–275
instruction types, 37–38
I/O architecture, 250
lack of conceptual integrity, 294
model, 25, 28
microprogrammed oeprating-system functions, 32
performance, 16
PERFORM instruction, 27
registers, 23, 48, 275, 294
semantic gap, 11–13
Implementation freedom, 295
Imprecise interrupt, 9
Index, 252, 254, 269
Index registers, 10, 12
Indirect addressing, 10
Information theory, 273, 279, 281
Input/output architecture, 249–272
Input/output processor, see Channel
Instruction frequencies, 274, 276–277, 279–285
Instruction-set optimization, 273, 274–279
Instruction trace, 180, 205, 206, 207, 210, 242
Interrupt, 10, 13
I/O device independence, 267, 270
IPL-language machine, 29
ISP language, 296
ISPL language, 28

Join operation, 257, 262
JOVIAL, 27, 28

KDF.9 machine, 27

Language-directed architecture, 25, 26–28, 31, 61, 151, 290, 292
Language distortions, 17–18, 23
Language-fragment method, 283
Language validation, 296–297
LARC computer, 10
Levels of abstraction, 3, 4
Lexical-level addressing, 50–53, 55, 67, 71, 84, 183, 278, 286
Linkage editing, 145, 210, 219

List, 103, 105
Literal operand, 117, 119, 161–162, 202, 278–279
L-Machine, 27
Locality of reference, 266–267, 288, 290
Logic error, 180, 182, 183

Machine description language, 83, 296
MARY language, 27
Mask register, 252–253
MDL, 296
Memory I/O, 268
Mental compilation, 296
Microdata 1621, 27
Microprogram, absence in SYMBOL, 146
  B1700, 28, 154
  and computer architecture, 6
  cost, 29
  data conversion, 42
  definition, 5
  generated by compilers, 32
  IBM S/370, 28
  Microdata 1621, 27
  processor advance, 9
Multics system, 267
Multiprocessing, 10, 99, 125, 145

One-level store, 267–269, 294, 296
ON-unit, 13, 103, 106, 114, 188
Op-code encoding, 161, 202, 279–285, 292, 295
Operating system, and
  application-directed architecture, 26, 32
  B1700, 152–153, 283, 286–287
  CAL, 188
  and computer architect, 7
  experimental architecture, 179
  Multics, 267
  SWARD machine, 188, 269–270
  SYMBOL system, 100, 136–140
Optimization, 32, 273–292
Orthogonality, 295

Pegasus computer, 10
Performance problems, 15–16
Physical input/output architecture, 5
Pipelining, 9
PL/I, ALIGNED attribute, 18
  binary arithmetic, 18
  and capability-based addressing, 54
  controlled storage, 13, 62, 66
  data conversion overhead, 42
  data structures, 188, 194
  decimal arithmetic, 17–18
  DEFINED attribute, 223
  EXTERNAL attribute, 224
  language-directed architectures, 27–28
  language study, 48
  lexical-level addressing, 50–53
  as machine description language, 296
  software errors, 14–15, 181–182, 183, 222
  and SPL, 103, 106
  and Student-PL machine, 297
  SUBSCRIPTRANGE, 15
  SWARD machine, 198, 199, 222, 224
  and S/370, 11–13
  variable-size data, 22
  VARYING attribute, 193
Pointer variable, 54, 181, 182
Procedure call, 12. See also Subroutine management
Processor architecture, 5, 9
Processor/memory bandwidth, 16, 40, 47, 81, 199, 273, 279, 280, 288, 289
Processor organization, 5, 9, 153–154, 295–296
Processor parallelism, 45
Programming errors, 14–15, 22, 39, 45, 50, 54, 81, 179–185, 222
Programming language I/O, 269
Program size, 16–17
Projection operation, 257, 263
Pseudovariable, 93, 104
Push-down stack, 13, 46, 63. See also Activation stack; Expression-evaluation stack

RAP, cell, 256–257, 264, 265
  controller, 255–256
  instruction format, 259–260
  instruction-set specification, 260–265
  mark bits, 258–260
  physical data representation, 258–259
  registers, 260, 262, 263–265
  relational data base, 256–257
  sample programs, 265–266
  set processor, 255–256
  virtual memory, 266–267
Raytheon AADC machine, 27

INDEX

Record, see Structures
Redundancy, 41–42, 81, 279–282, 284, 287–288, 290
Reference variable, see Pointer variable
Register, 23, 47
Register addressing, 23
Register-oriented instruction set, 46, 48, 57, 187
Relation, 257, 258–260
Relational associative processor, see RAP
Relational data base, 255, 256–258
Response array, 253–254, 258
Reverse Polish notation, 28, 46, 48, 70, 97
Rice Research Computer, 27
RPG, 152, 155, 156

Scope of names, 12, 106, 182, 209, 236
SDL language, 152
Self-defining data objects, see Descriptor
Self-identifying storage, see Tagged storage
Semantic error, 180–185, 186
Semantic gap, 11–19, 24, 25, 146, 151–152, 222, 249–250, 293
Send /receive mechanism, 270
Set operation, 253, 255, 257, 263
Shift register, 254, 255, 256
Single-level store, see One-level store
S-language, 153
SNOBOL, 27
Software architecture, 4
Software error studies, 181–182, 222
Software reliability, 14–15, 27, 179–185, 222
Source entropy, 279
Source /sink I /O, 268–270
Source uncertainty, 279
Space Programming Language, 28
Stack, see Push-down stack
Stack-oriented instruction set, see Expression-evaluation stack
Staran, 255
Static instruction frequencies, 274, 276–277, 280
Storage hierarchy, 267, 268
Storage object, 53–54, 185, 209
Storage protection, 54, 55, 186
Storage-to-storage addressing, 48–49, 57, 156, 186
Strings, 12, 188, 192–193
Structures, 12, 20, 43, 188, 194–195

Student-PL machine, comparison to other architectures, 80–81, 123, 183–184, 215, 279
 formal description, 296
 instruction-set specification, 83–94
 lack of branch instruction, 73–77
 language, 61–63
 language-directed architecture, 27, 61
 memories, 63–68
 optimization, 276–279, 291
 and PL /I, 297
 problems, 81
 sample programs, 69–81
Subroutine management, B1700, 156, 173
 definition, 49–50
S /370 PL /I Optimizing Compiler, 80
Student-PL machine, 66–67, 77–79, 91–93
SWARD machine, 187, 188
SYMBOL system, 110, 121–123
Supplemental instruction sets, 197–198, 199–200
SWARD machine, address optimization, 287–290
 address space, 194, 200–201, 202
 application-directed architecture, 33
 automatic storage die, 200–201, 209, 211, 212
 capability-based addressing, 185–186, 193–194, 268–270
 cell, 190
 cell address, 194, 199, 200, 202, 288–290
 comparison to other architectures, 215
 constant storage die, 201, 204, 218
 entry-point zero, 200, 205–207
 exterior storage die, 200
 fault handling, 188, 200, 203–207, 240–241, 246, 270
 formal description, 296
 input /output, 223, 268–270
 instruction format, 202–203, 283–286
 instruction-set specification, 207–210, 225–242
 internal storage objects, 244–246
 language-directed architecture, 27–28
 logical address, 185, 193–194, 204, 214, 219, 268–270
 module, 185, 194, 198–202, 205, 239, 246
 nested tags, 195–197, 225
 op-code encoding, 283–286, 292

operand address, 202–203
primitive cells, 191–194, 196, 197
problems, 223, 297
sample programs, 210–222
send/receive mechanism, 270
software-reliability-directed architecture, 179–183
static storage die, 201, 218
storage die, 187, 200–201
storage object, 185, 209, 268
structure cell, 194–195, 197
subroutine management, 187, 209, 219
supplemental instruction set, 197–198, 199–200
tagged storage, 186, 187, 188, 190
token, 190
variable-size addresses, 199, 242–244
SYMBOL system, application-directed architecture, 33
  central processor, 100–101, 110–123, 136, 138–140, 141–144
  channel controller, 99, 101–102, 145
  comparison to other architectures, 123, 183–184, 215
  disk controller, 100, 137, 145
  high-level-language architecture, 29
  instruction format, 117
  interface processor, 99, 101–102, 110, 138–139, 145
  LIMIT, 104, 120, 143
  logical storage, 128–130, 145
  main bus, 101, 125–127
  maintenance processor, 100, 145
  memory controller, 100–102, 126–136, 145
  memory reclaimer, 100–101, 127, 133–134, 136
  name table, 102, 112–117, 144, 159
  paging, 100–101
  subroutine management, 121–123
  SYMBOL Programming Language, 97, 103–107, 140–141
  system supervisor, 100–102, 127, 136–140
  tagged storage, 39, 112
  translator, 100–102, 112, 128, 138–139, 144–145
  virtual storage, 128, 130–132, 133, 137–139
Syntax-driven machine, 29

System architecture, 3–4
S/370, see IBM S/360-370

Tagged storage, B1700, 156
  and capability-based addressing, 54
  and conceptual integrity, 294
  debugging, 180
  definition, 37–43
  nested tags, 195–197
  redundancy remover, 41–42, 290
  storage requirements, 40–42, 290
  Student-PL machine, 63–66
  and subroutine management, 50, 187
  SWARD machine, 186, 188, 191–196
  SYMBOL system, 110, 112
Technology independence, 295–296
Telecommunications access method, 251
Testing, 180, 183, 222
TPL language, 27
Trap bits, 38, 40
Tuple, 257, 258–259
Type-A high-level-language architecture, 28–29, 31, 61
Type-B high-level-language architecture, 29, 31, 110
Type-C high-level-language architecture, 29–31
Typed storage, see Tagged storage
Typeless language, 103

Undefinded value, 14, 38–39, 63, 73, 81, 180, 181, 187, 191, 204
Unified view of storage, 250
Uniquely decipherable op-code, 281
Unique-name addressing, see Capability-based addressing
Univac 1103, 10

Variable-size operation code, see Op-code encoding
Variable-size storage words, 55, 110, 156, 158, 186, 199, 242–244
Virtual storage, 10, 17, 128, 130–133, 250, 266–267
Von Neumann architecture, 19–21, 24, 30, 37, 151–152, 183, 290, 295, 296–297

Working set, 17, 274

XPL language, 290